AF335609

EMERGING AVIAN DISEASE

EMERGING AVIAN DISEASE

Ellen Paul, *Editor*

Studies in Avian Biology No. 42

A PUBLICATION OF THE COOPER ORNITHOLOGICAL SOCIETY

University of California Press
Berkeley   Los Angeles   London

University of California Press, one of the most distinguished university presses in the United States, enriches lives around the world by advancing scholarship in the humanities, social sciences, and natural sciences. Its activities are supported by the UC Press Foundation and by philanthropic contributions from individuals and institutions. For more information, visit www.ucpress.edu.

Studies in Avian Biology No. 42

For e-book, see the UC Press website.

University of California Press
Berkeley and Los Angeles, California

University of California Press, Ltd.
London, England

Library of Congress Cataloging-in-Publication Data

Emerging avian disease / Ellen Paul, volume editor. — 1st ed.
    p. cm.
    Includes bibliographical references and index.
    ISBN 978-0-520-27237-8 (hardback)
    1. Birds—Diseases. 2. Bird populations. 3. Ornithology.
    I. Paul, Ellen 1955– II. Symposium on Avian Disease
(2006 : Veracruz, Mexico)
    SF994.E64 2012
    636.5'089—dc23

                                        2011052142

            19 18 17 16 15 14 13 12
            10 9 8 7 6 5 4 3 2 1

The paper used in this publication meets the minimum requirements of ANSI/NISO Z39.48-1992 (R 2002)(*Permanence of Paper*).

Cover image: Razorbill chick (*Alca torda*) with ticks, Labrador, Canada, 2005. Photo by Jennifer Lavers.

PERMISSION TO COPY

CONTENTS

CONTRIBUTORS

ROSEMARY K. BARRACLOUGH
Disease and Vector Research Group
Institute of Natural Sciences
Massey University, Albany Campus
Private Bag 102-904, NSMC
Auckland, New Zealand
r.k.barraclough@massey.ac.nz

SABIR BIN MUZAFFAR
Department of Biology Faculty of Science
United Arab Emirates University P.O. Box 17551
Al Ain, United Arab Emirates
s_muzaffar@uaeu.ac.ae

DIANNE H. BRUNTON
Ecology & Conservation Research Group
Institute of Natural Sciences
Massey University, Albany Campus
Private Bag 102-904, NSMC
Auckland, New Zealand
d.h.brunton@massey.ac.nz

DONALD F. CACCAMISE
New Mexico State University
Agricultural Experimental Station
P.O. Box 30003, MSC 3BF
Las Cruces, NM 88003, USA
dcaccami@nmsu.edu

BRUCE K. CAHILL
Vector-borne Disease Laboratory
Maine Medical Center Research Institute
75 John Roberts Road, Suite 9B
South Portland, ME 04106, USA
cahilb@mail.mmc.org

TANEAL M. COPE
Department of Zoology
University of Melbourne
Parkville Campus
Melbourne 3010, Australia
taneal.cop@gmail.com

REBECCA CREAMER
Department of Entomology
Plant Pathology and Weed Science
New Mexico State University
Skeen Hall Room N141
P.O. Box 30003, MSC 3BE
Las Cruces, NM 88003, USA
creamer@nmsu.edu

ROBERT J. DUSEK
USGS National Wildlife Health Center
6006 Schroeder Road
Madison, WI 53711, USA
rdusek@usgs.gov

JEANNE M. FAIR
Los Alamos National Laboratory
Biosecurity and Public Health
P.O. Box 1663
Los Alamos, NM 87545, USA
jmfair@lanl.gov

ERIK K. HOFMEISTER
USGS Fort Collins Science Center
2150 Centre Avenue, Building C
Fort Collins, CO 80526, USA
ehofmeister@usgs.gov

WILLIAM M. IKO
USGS National Wildlife Health Center
6006 Schroeder Road
Madison, WI 53711, USA
kob@usgs.gov

IAN L. JONES
Department of Biology
Memorial University of Newfoundland
St. John's, NL A1B 3X9, Canada
iljones@mun.ca

ELEANOR H. LACOMBE
Vector-borne Disease Laboratory
Maine Medical Center Research Institute
75 John Roberts Road, Suite 9B
South Portland, ME 04106, USA
elacombe@midcoast.com

JENNIFER LAVERS
Department of Biology
Memorial University of Newfoundland
St. John's, NL A1B 3X9, Canada
jennifer.lavers@csiro.au

CHARLES B. LUBELCZYK
Vector-borne Disease Laboratory
Maine Medical Center Research Institute
75 John Roberts Road, Suite 9B
South Portland, ME 04106, USA
lubelc@mmc.org

BABETTA L. MARRONE
Los Alamos National Laboratory
Biosecurity and Public Health
P.O. Box 1663
Los Alamos, NM 87545, USA
blm@lanl.gov

ROBERT G. MCLEAN
National Wildlife Research Center
WS/APHIS/USDA
4101 Laporte Avenue
Fort Collins, CO 80521-2154, USA
robert.g.mclean@aphis.usda.gov

PATRICIA G. PARKER
University of Missouri–St.Louis
Whitney R. Harris World Ecology Center
Department of Biology
One University Boulevard
St. Louis, MO 63121, USA
(Current address: Saint Louis Zoo
One Government Drive
St. Louis, MO 63110, USA
pparker@umsl.edu)

ELLEN PAUL
Ornithological Council
5107 Sentinel Drive
Bethesda, MD 20816, USA
ellen.paul@verizon.net

MICHAEL A. PEIRCE
MP International Consultancy
6 Normandale House
Normandale Bexhill-on-Sea
East Sussex, TN39 3NZ, United Kingdom
mapeirce.mpic@btinernet.com

A. TOWNSEND PETERSON
Natural History Museum and Biodiversity
Research Center
University of Kansas
Lawrence, KS 66045, USA
town@ku.edu

ORNITHOLOGICAL COUNCIL
5107 Sentinel Drive
Bethesda, MD 20816, USA
ellen.paul@verizon.net

PETER W. RAND
Vector-borne Disease Laboratory
Maine Medical Center Research Institute
75 John Roberts Road, Suite 9B
South Portland, ME 04106, USA
randp@mmc.org

MARTA REMMENGA
National Surveillance Unit
USDA/APHIS/VS/CEAH
2150 Centre Avenue, Building B,
Mail Stop 2E6
Fort Collins, CO 80526, USA
marta.d.remmenga@aphis.usda.gov

ROBERT E. RICKLEFS
University of Missouri–St. Louis
Whitney R. Harris World Ecology Center
Department of Biology
One University Boulevard
St. Louis, MO 63121, USA
ricklefs@umsl.edu

DIEGO SANTIAGO-ALARCON
University of Missouri–St.Louis
Whitney R. Harris World Ecology Center
Department of Biology
One University Boulevard
St. Louis, MO 63121, USA
onca77@yahoo.com

YULIN SHOU
Los Alamos National Laboratory
Biosecurity and Public Health
P.O. Box 1663
Los Alamos, NM 87545, USA
shou@lanl.gov

ROBERT P. SMITH, JR.
Vector-borne Disease Laboratory
Maine Medical Center Research Institute
75 John Roberts Road, Suite 9B
South Portland, ME 04106, USA
smithr@mmc.org

KIRSTEN J. TAYLOR-MCCABE
Los Alamos National Laboratory
Biosecurity and Public Health
P.O. Box 1663
Los Alamos, NM 87545, USA
kjmccab@lanl.gov

HOLLY B. VUONG
Department of Fishery and Wildlife Sciences
New Mexico State University
2980 South Espina, Knox Hall 132
P.O. Box 30003, MSC 4901
Las Cruces, NM 88003, USA
hvuong@eden.rutgers.edu

Wildlife biologists once believed that mortality from diseases in wildlife populations was compensatory and, thus, did not affect populations. Instead, they focused on other sources of mortality when conducting population studies. Many avian biologists also held that belief because most disease outbreaks were sporadic and self-limiting, resulting in minor losses. Even then, however, there were localized threats from bird malaria and avian pox virus, for example, to the endangered native bird populations of Hawaii and some large, very localized mortality events among waterbirds from diseases such as avian botulism, avian cholera, and Newcastle disease (Friend et al. 2001). This prevailing perspective changed during the last few decades due to the invasion, emergence, or reemergence of some major diseases of free-ranging wild birds. Avian diseases have increased in frequency of occurrence, prominence, and geographical distribution and resulted in frequently occurring mortality events with major losses of birds of a wide variety of species. Increased attention also resulted from the recognition that 175 new human diseases have emerged in the last two decades, nearly 75% of them caused by zoonotic disease agents that are transmitted among wild or domestic animals and humans (Daszak et al. 2000, Gibbs 2005). Additionally, a troubling number of diseases directly affecting wildlife species and populations emerged (Friend et al. 2001, McLean 2007).

Recent avian examples include: the rapid expansion and spread of conjunctivitis caused by *Mycoplasma gallisepticum* from poultry to House Finches (*Carpodacus mexicanus*) from the east coast of the United States in 1994 to the Mississippi River (Hartup et al. 2001, Dhondt et al. 2005); a major die-off of American White Pelicans (*Pelacanus erythrorhynchos*) and Brown Pelicans (*P. occidentalis*) on the Salton Sea in California in 1996 from a unique type C botulism (Rocke et al. 2004); and the dramatic mortality of North American birds from West Nile virus (WNV), especially corvids, American White Pelicans, and Greater Sage-Grouse (*Centrocercus urophasianus*), following its introduction to North America in 1999 (McLean 2002, Naugle et al. 2004, Rocke et al. 2005, Clark et al. 2006, LaDeau et al. 2007). The establishment and subsequent spread of WNV across the continent within five years of introduction (McLean 2006) was a primary example of an emerging zoonotic disease affecting human, domestic animal, and wildlife populations. The potential threat of direct transmission of WNV and other zoonotic diseases of birds to bird banders and ornithologists led to the development of guidelines and information for wild bird handlers to protect themselves and prevent infection (Ornithological Council 2010).

This volume brings together some important information on emerging diseases of wild birds, with topics ranging from how cellular mechanisms of the immune system determine susceptibility to pathogens to the global movement of ticks and bacteria by colonial seabirds. WNV is a central focus, starting with the development of laboratory methods to immunophenotype lymphocytes that determine natural variability in immunocompetence. Two other studies included here on WNV document bird species susceptibility and exposure rates in the southwestern United States. Despite nationally reported mortality from WNV in a number of raptor species in the United States (Nemeth et al. 2006, 2007), including American Kestrels (*Falco sparverius*), a population study of this species in Colorado found nearly all of the adult birds were immune the year following the 2003 Colorado outbreak. The field study suggests that American Kestrels, at least in Colorado, are not as susceptible to WNV as previously thought and that relatively few birds in the study population died from the infection. Another study of WNV transmission in southern New Mexico revealed differential infection rates among species and habitats. The desert and riparian habitats had higher bird diversity and lower seroprevalence to WNV compared to urban and agriculture habitats.

Parasitic disease and vector transmission is another focus of this volume. For example, haemosporidian parasites (*Haemoproteus* sp.) in endemic Galápagos Doves (*Zenaida galapagoensis*), illustrate the negative relationship between host biodiversity and disease prevalence. Host composition significantly affected transmission rates, although other factors affecting disease transmission, particularly vector population dynamics, were mentioned as contributing factors; these were not directly investigated.

The emergence and spread of avian malaria (*Plasmodium* sp.) is revealed within an endemic population of the New Zealand honeyeater, the Bellbird (*Anthornis melanura*). A relatively high prevalence of malaria infections was detected without any of the deleterious effects previously observed in the native bird species of Hawaii. Global spread of ticks and *Borrelia garinii* (a vector and a spirochete), which is a human pathogen of colonial seabirds, and the potential global spread of avian influenza viruses by waterbirds illustrate connectivity through the natural movements of migratory birds and anthropogenic transportation of pathogens.

A study of seabirds in different nesting colonies in the northwest Atlantic discovered the recent arrival of *B. garinii*—a spirochete. Previous records of *B. garinii* indicated that the pathogen was endemic in seabird colonies in the greater North Atlantic since at least the early 1990s and subsequently spread into the northwest Atlantic colonies. Prevalence of spirochete infections in ticks, *Ixodes uriae*, removed from seabirds and soil from nesting colonies varied among years and avian host species, with prevalence rates ranging up to 37.5%. Atlantic Puffins (*Fratercula arctica*), among other seabird species, seem to be suitable reservoirs.

A third focus of this volume is viral infection. A unique study examined the movement patterns of selected migratory waterfowl species and suggested how the highly pathogenic H5N1 influenza virus might be introduced. Established virus foci in Europe and Asia spread to North America, then from northern bird breeding regions to the wintering regions farther south. A forecast for H5N1 dispersal via migratory bird movements in North America emphasizes the migration pattern of six species of arctic-breeding Anseriformes along the coasts of North America and along the central Mississippi River system.

Three major viral pandemics with animal origins occurred during the first decade of this century. Pandemics from SARS, H5N1 HPAI, and H1N1 had high economic costs and major political consequences. Many factors are responsible for dramatic emergences of these zoonotic diseases, including unprecedented worldwide human population growth; global wildlife trade; changes in food quality and abundance; and increases in the frequency and speed of international travel to transport people, food, products, wildlife, and the vectors, diseases, and pathogens that accompany them. Emerging threats are likely to increase in frequency and magnitude, and continued global surveillance and research on infectious zoonotic diseases are critical. With the continuing threat of diseases to free-living birds globally and the potential influence of climate change on diseases, synthetic reviews on avian diseases should continue to be a high priority for ornithological research.

ROBERT G. MCLEAN
National Wildlife Research Center,
Fort Collins, Colorado

Literature Cited

Clark, L., J. Hall, R. G. McLean, M. Dunbar, K. Klenk, R. Bowen, and C. A. Smeraski. 2006. Susceptibility of Greater Sage-Grouse to experimental infection with West Nile virus. Journal of Wildlife Diseases 42:14–22.

Daszak, P., A. A. Cunningham, and A. D. Hyatt. 2000. Emerging infectious diseases of wildlife: threats to biodiversity and human health. Science 287:443.

Dhondt, A., S. Altizer, E. G. Cooch, A. K. Davis, A. Dobson, M. J. L. Driscoll , B. K. Hartup, D. M. Hawley,W. M. Hochachka, P. R. Hosseini, C. S. Jennelle, G. V. Kollias, D. H. Ley, E. C.H. Swarthout, and K.V. Sydenstricker. 2005. Dynamics of a novel pathogen in an avian host: *Mycoplasmal conjunctivitis* in House Finches. Acta Tropica 94:77–93.

Friend, M., R. G. McLean, and F. J. Dein. 2001. Disease emergence in birds: challenges for the twenty-first century. The Auk 118:290–303.

Gibbs, E. P. J. 2005. Emerging zoonotic epidemics in the interconnected global community. Veterinary Record 157:673–679.

Hartup, B. K., J. M. Bickal, A. A. Dhondt, D. H. Ley, and G. V. Kollias. 2001. Dynamics of conjunctivitis and *Mycoplasma gallisepticum*. The Auk 118:327–333.

LaDeau S. A., A. M. Kilpatrick, and P. P. Marra. 2007. West Nile virus emergence and large-scale declines of North American bird populations. Nature 447:710–714.

McLean, R. G. 2002. West Nile virus: a threat to North American avian species. Transactions of the North American Wildlife and Natural Resource Conference 67:62–74.

McLean, R. G. 2006. West Nile virus in North American birds. Ornithological Monographs 60:44–64.

McLean, R. G. 2007. Introduction and emergence of wildlife diseases in North America. Pp. 261–278 in T. E. Fulbright and D. G. Hewitt (editors), Wildlife science: linking ecological theory and management applications. CRC Press, Boca Raton, FL.

Naugle, D. E., C. L. Aldridge, B. L. Walker, T. E.Cornish, B. J. Moynahan, M. J. Holloran, K. Brown, G. D. Johnson, E. T. Chmidtmann, R. T. Mayer, C. Y. Kato, M. R. Matchett, T. J. Christiansen, W. E. Cook, T. Creekmore, R. D. Falise, E. T. Rinkes, and M. S. Boyce. 2004. West Nile virus: pending crisis for Greater Sage-Grouse. Ecology Letters 7:704–713.

Nemeth, N., D. Gould, R. Bowen, and N. Komar. 2006. Natural and experimental West Nile virus infection in five raptor species. Journal of Wildlife Diseases 42:1–13.

Nemeth, N., G. Kratz, E. Edwards, J. Scherpelz, R. Bowen, and N. Komar. 2007. Surveillance for West Nile virus in clinic-admitted raptors, Colorado. Emerging Infectious Diseases 13:305–307.

Ornithological Council. 2010. West Nile virus, highly pathogenic avian influenza H5N1, and other zoonotic diseases: what ornithologists and bird banders should know. <http://www.nmnh.si.edu/BIRDNET/documents/WNV&H5N1-FactSheet.pdf>.

Rocke, T., K. Converse, C. Meteyer, and B. McLean. 2005. The impact of disease in American White Pelicans in North America. Waterbirds 28 (Special Publication 1):87–94.

Rocke, T. E., P. Nol, C. Pelliza, and K. Sturm. 2004. Type C botulism in pelicans and other fish-eating birds at the Salton Sea, California, 1994–2001. Studies in Avian Biology 27:136–140.

PART ONE

Environmental and Behavioral Aspects of Transmission

Ecological Associations of West Nile Virus and Avian Hosts in an Arid Environment

Holly B. Vuong, Donald F. Caccamise, Marta Remmenga, and Rebecca Creamer

Abstract. We evaluated disease associations of West Nile virus (WNV) with avian hosts in four key habitats of southern New Mexico (agriculture, desert, riparian, and urban). Our goal was to examine the role of avian life history traits in transmission of WNV and to evaluate possible mechanisms to explain differences in seroprevalence among avian communities. Seroprevalence was highest in Summer Tanagers (*Piranga rubra*, 39%) and American Robins (*Turdus migratorius*, 33%). Serosurveys of bird communities indicated differences among habitats, age, and resident status. Urban and agricultural habitats had higher seroprevalence than desert and riparian habitats. After-hatch-year birds had higher seroprevalence than hatch-year birds. Seroprevalence in permanent resident and local breeding species were higher than migrants and winter residents. Males had higher seroprevalence in 2004, while females were higher in 2005. Analyses among communities indicated negative relationships between seroprevalence and avian species diversity and richness. Desert and riparian habitats had higher diversity and lower seroprevalence compared to urban and agriculture. This study revealed associations between WNV and avian life history traits, providing insights into mechanisms of transmission in the arid Southwest. In addition, we found relationships between complexity of avian host communities (e.g., species diversity, species richness) and patterns in seroprevalence of WNV in avian host species.

Key Words: avian community, desert, habitat, life history, seroprevalence, southern New Mexico, West Nile virus, WNV.

Asociaciones Ecológicas del Virus del Nilo Occidental y sus Aves Hospedadoras en un Ambiente Árido

Resumen. Evaluamos las asociaciones de la enfermedad del virus del Nilo Occidental (VNO) con sus aves hospedadoras en cuatro hábitats clave del Sur de Nuevo México (agrícola, desértico, ripario y urbano). Nuestro objetivo fue examinar el papel que juegan las características de las historias de vida de las aves en la transmisión del VNO, y evaluar los posibles mecanismos que expliquen las diferencias de la seroprevalencia entre las distintas comunidades de aves. La seroprevalencia más alta se observó en la Tángara Roja migratoria (*Piranga rubra*, 39%), y en el Zorzal Pechirrojo (*Turdus migratorius*, 33%). El estudio serológico de las comunidades de aves indicó que existen diferencias entre hábitats, edad, y estatus de residencia. Las aves de hábitats urbanos y agrícolas presentaron una seroprevalencia mayor que las

Vuong, H. B., D. F. Caccamise, M. Remmenga, and R. Creamer. 2012. Ecological associations of West Nile virus and avian hosts in an arid environment. Pp. 3–22 *in* E. Paul (editor). Emerging avian disease. Studies in Avian Biology (vol. 42), University of California Press, Berkeley, CA.

de hábitats desérticos y riparios. La seroprevalencia fue mayor en aves adultas que en juveniles. La seroprevalencia de las especies residentes y de las especies que se reproducen localmente fue mayor que la de las especies migratorias y las residentes invernales. Los machos mostraron una mayor seroprevalencia en 2004, mientras que las hembras tuvieron una mayor seroprevalencia en 2005. Los análisis para las diferentes comunidades indicaron que existe una relación negativa entre la seroprevalencia y la diversidad y riqueza de especies de aves. Los hábitats desérticos y riparios presentaron una mayor diversidad y menor seroprevalencia que los hábitats urbanos y agrícolas.

Este estudio reveló la existencia de asociaciones entre el VNO y las características de las historias de vida de las aves, proporcionando información sobre los mecanismos de transmisión en el ambiente árido del Suroeste de los Estado Unidos. Además, encontramos relaciones entre la complejidad de las comunidades de aves hospedadoras (por ejemplo, la diversidad y riqueza de especies) y los patrones de seroprevalencia del VNO en las especies de aves infectadas.

Palabras Clave: comunidad aviar, desierto, hábitat, historia de vida, meridional, Nuevo México, seroprevalencia, Virus del Nilo Occidental, VNO.

I n the US, 326 bird species in 57 families and 23 orders have tested positive for West Nile virus (WNV) infection since the virus was first detected in New York City in 1999 (CDC 2009). As an emerging disease in the U.S., WNV expanded across the Western Hemisphere, developing new host–virus associations along the way. Initial studies in the U.S. documented the spread of WNV and were based mainly on surveillance for dead birds (Bernard et al. 2001; Eidson et al. 2001a, 2001b). Such studies provided information on the impact of infection on individual species, but they failed to examine the ecological dynamics of WNV in relation to host communities. More recently, research efforts have focused on the role of individual species in the associations between WNV and avian host communities (Beveroth et al. 2006, Ezenwa et al. 2006, Marshall et al. 2006, Harris and Sleeman 2007, Wilcox et al. 2007). In addition, reservoir competency studies under laboratory conditions provided information on the susceptibility of individual species to WNV and their potential role in the transmission cycle (Komar et al. 2003). Studies based on capture of free-living birds have provided information on exposure to the virus under natural conditions (Ringia et al. 2004, Beveroth et al. 2006, Marshall et al. 2006).

The life history traits of birds, including residency status, habitat, age, and sex, can influence disease prevalence. Resident birds have been shown to have higher seroprevalence for WNV (Ringia et al. 2004, Beveroth et al. 2006, Gibbs et al. 2006) and other mosquito-borne encephalitic viruses (Crans et al. 1994, Goddard et al. 2002,

Reisen et al. 2004b) that may be due to the longer seasonal overlap of residents with mosquito vectors. Nonetheless, migratory birds may be key players in transporting the virus to new areas (Rappole et al. 2000). Studies of habitat effects for WNV have focused primarily on urban versus rural areas (Taylor et al. 1956, McIntosh and Jupp 1982, Tsai et al. 1998, Ringia et al. 2004). These past field studies demonstrated that seroprevalence in humans and avian hosts and WNV prevalence in mosquitoes tend to be higher in urban habitats.

Several similar arboviruses, such as eastern equine encephalitis (EEE; Crans et al. 1994), St. Louis encephalitis (SLE; Gruwell et al. 2000, Reisen et al. 2000), and western equine encephalitis (WEE; Reisen et al. 2000), along with WNV (Ringia et al. 2004, Beveroth et al. 2006, Gibbs et al. 2006), have shown differences in seroprevalence between age classes, with after-hatch-year birds (AHY) generally showing higher rates than hatch-year (HY) birds. Nonetheless, HY birds are important in the transmission cycle of arboviruses because they are naïve, susceptible hosts when they enter the population, and could serve to amplify disease transmission and help maintain the pathogen for extended periods (Reed and Crans 1998, Hamer et al. 2008). A sex bias in seroprevalence might provide new insights into the ecology of host–pathogen interactions. The first study to show sexual differences in seroprevalence reported that female Northern Cardinals (*Cardinalis cardinalis*) in Ohio had higher WNV seroprevalence than males (Marshall et al. 2006). The authors suggested that this pattern appeared because females may be more exposed to mosquitoes during incubation. Other studies on WNV (Ringia et al. 2004, Beveroth et al. 2006, Gibbs et al. 2006) and similar arboviruses (Crans et al. 1994; Gruwell et al. 2000; Reisen et al. 2000, 2001, 2005) have not found differences in seroprevalence between sexes.

The importance of avian hosts in disease transmission can vary in relation to both community complexity and host responses to variations in local ecological conditions. When few primary reservoirs are associated with host communities, disease risk may be reduced. For example, risk of Lyme disease in humans is inversely related to complexity of host communities (Ostfeld and Keesing 2000, LoGuidice et al. 2003). The authors suggested that risk might decline if the primary competent host is relatively less abundant in more complex communities. A similar relationship was also shown with hantavirus (Mills 2006) and WNV (Ezenwa et al. 2006).

As WNV spread into New Mexico, we expected that differences in life history traits among avian hosts would result in variations in seroprevalence. Furthermore, we predicted that seroprevalence would vary among habitat types based on differences in the composition of avian host communities. Therefore, the goals of this study were to (1) evaluate the variation in disease associations of avian hosts among four key habitats of southern New Mexico (riparian, agriculture, urban, and desert), (2) examine the role of avian life history traits in the transmission of WNV, and (3) evaluate the relationships between complexity of avian host communities and patterns in seroprevalence.

METHODS

Study Area

The study area lies within the arid Chihuahuan Desert of southern New Mexico, in Doña Ana County (Fig. 1.1). With only four mosquito species making up approximately 85% of the local mosquito community, this area offers a unique opportunity to study WNV where host–vector interactions are relatively simple (Pitzer et al. 2009). The Chihuahuan Desert supports a large diversity of birds that pass through the riparian corridor of the Rio Grande River during fall and spring migrations. The arid climate in the region yields an average of 23.5 cm of rain per year, with 55% falling between July and September, mainly in the form of localized thunderstorms (Western Regional Climate Center 2009). The landcover is composed largely of upland desert mesas dominated by typical desert-shrub vegetation (Dick-Peddie 1999). The Rio Grande River valley bisects Doña Ana County from northwest to southeast and encompasses urbanized areas along with agricultural lands and highly modified riparian habitats adjacent to the river. We located three study sites within each of the four primary cover types in the area, including desert, urban, agricultural, and riparian, for a total of 12 study sites.

Desert sites were located on mesas at 3–25 km from the Rio Grande River. We chose sites with impoundment ponds for cattle watering or natural playas that provide aquatic habitats suitable for mosquito breeding when filled with summer rain. Urban sites were located within the city limits of Las Cruces. We selected residential areas that provided habitats attractive to birds. Vegetation in the residential sites is sustained by irrigation consisting of combinations of drip systems, flood irrigation, and hand watering. All agricultural crops in the Rio Grande valley are irrigated, mainly by flood irrigation. Water arrives to fields through a network of delivery canals and is drained from fields through a system of return canals that lead back to the river. Delivery canals are maintained vegetation free, but return canals have standing water throughout the year and generally support dense vegetation along the banks that is heavily used by birds (Thompson et al. 1994). The lower Rio Grande is channelized so riparian habitats are sparse, occurring where restored wetlands and narrow vegetated patches are adjacent to the river.

Avian Sampling

The field season ran from March through mid-October in 2004 and 2005. Two field crews sampled each of 12 field sites on an 8–10-day rotating schedule. We captured birds using mist nets (12 m × 2.5 m, mesh size 30 mm and 61 mm; AFO Mist Nets, Manomet, Inc., Manomet, MA), as described by Ralph et al. (1993). We set 5–10 nets per site about 30 min before sunrise and checked them every 20 min for approximately 3–4 hr. We recorded species, weight, sex, and age when possible for each captured bird (Pyle 1997, Sibley 2000) and applied a U.S. Geological Survey aluminum leg band.

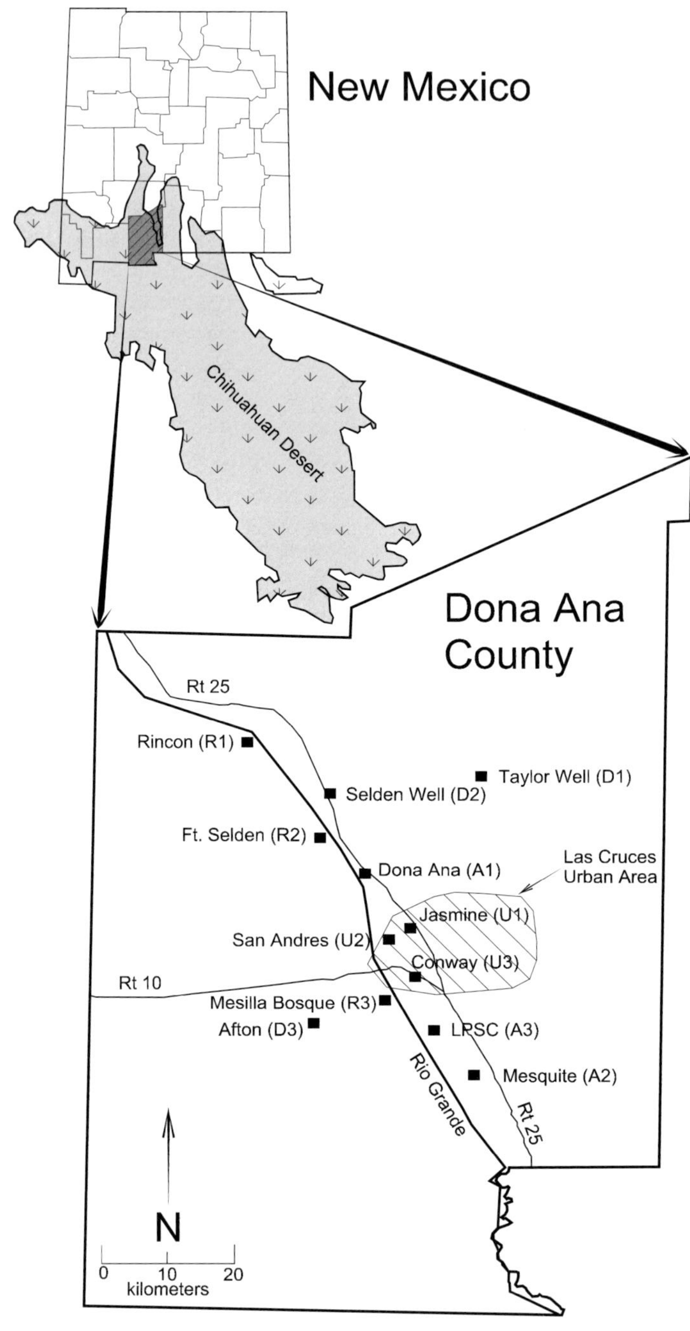

Figure 1.1. Map of study area showing the 12 field sites in Doña Ana County, New Mexico. Inset: New Mexico state map with the Chihuahuan Desert (overlay) extending from southern New Mexico into central Mexico.

We obtained a blood sample from each bird to test for WNV antibodies. For birds >10 g, we used jugular or brachial venipuncture (29 gauge needle) to obtain at least 100 µl of blood. We placed the blood samples into 2-ml microcentrifuge tubes containing 900 µl of 5% bovine serum albumin (BSA) in buffer (PBS containing 0.05% Tween 20) to provide a 1:10 blood dilution. For birds <10 g we used brachial venipuncture by lancet and absorbed blood directly onto filter paper strips (Whatman® 1 Qualitative, Newark, NJ). We kept the blood samples cool while in the field and during transport back to the laboratory. We centrifuged the blood samples at 530 g for 10 min, transferred the plasma to newly labeled microcentrifuge tubes, and stored them at $-20°C$ until testing. We eluted the blood samples from the filter paper strips overnight using 5% BSA buffer at 4°C and centrifuged the samples the next day at 530 g for 10 min. The eluted sera were then transferred into newly labeled tubes and stored at $-20°C$ until testing.

Serological Assays

We used a blocking ELISA (enzyme-linked immunosorbent assay) method described by Jozan et al. (2003) to detect the presence of WNV antibodies. The monoclonal antibody (MAb 3.1112G) used in the ELISA is highly specific to the NS1 epitope of WNV, showing essentially no cross reactivity with SLE (Jozan et al. 2003). We initially tested all samples at 1:20 dilution in Immulon® 2HB plates (96 well, flat bottom Microtiter® Plates; Thermo Labsystems, Franklin, MA). Plates were read by Emax Precision Microplate Reader (Molecular Devices Corporation, Sunnyvale, CA) using program SoftMax Pro (ver. 3.1.1) at 405 nm wavelength to obtain optical density (OD) values. Dilutions of the antigen and monoclonal (generally 1:2,000 and 1:2,500, respectively) were previously determined by titration.

We estimated antibody titer by calculating the inhibition level (IL) for the antibodies by:

$$IL = 100 - \left[\left(\frac{TS - B}{CS - B} \right) \times 100 \right]$$

where TS is the OD of test serum, B is the OD of the background for each test serum, and CS is the OD of negative control. IL provides a measure of the relative number of binding sites blocked by the antibodies and therefore unavailable to the monoclonal antibodies. When we obtained an IL >45% for our sample at 1:20 dilution, we reconfirmed by further titration (up to 1:80) before categorizing the sample as positive (Jozan et al. 2003).

Statistical Analyses

We grouped each species into one of four categories according to residency status. Permanent residents occur locally year round. Breeding birds spend their breeding period in the area but are absent during the winter. Migrants pass through during migration. Wintering birds spend their winters in southern New Mexico but do not breed locally (Sibley 2000).

We performed logistic regressions with the Genmod procedure in program SAS (ver. 9.1; SAS Institute, Cary, NC) using the binomial distribution and the logit link function to model the probability of each bird being positive for WNV antibodies as a function of habitat, month of capture, age, sex, or residency status. Not all levels of combination were included in this analysis because of too few data points to cover all levels of the interaction. However, logistic regressions on smaller interaction effects were conducted separately. Differences between pairs of proportions were tested with a chi-square test performed by the least square means statement in the Genmod procedure of SAS. We combined samples from March with April and September with October due to low numbers of captures at the start and end of field seasons. Individuals of undetermined age and sex were removed from analyses that included age and sex. Recaptured birds that were positive were only used once in the analyses, but every recapture prior to becoming seropositive was used in the analyses. To determine the importance of HY birds as a source population for virus cycling, we also examined temporal changes of seroprevalence in AHY and HY birds across the seasons.

We used the Shannon–Wiener diversity index to calculate avian diversity for all 12 study sites in each year, using $H' = -\Sigma p_i \ln p_i$ where p_i is the proportion of individuals in the ith species (Magurran 1998). We calculated evenness following Magurran (1998), using $J' = H'/H_{max}$, where H_{max} is the natural log of the species richness, and richness (S) equals the number of species against seroprevalence using the Mixed procedure in SAS (SAS software, version 9.1 of the SAS Systems for Windows). Multiple analyses inflated our type I error rate, and we corrected our analyses for number of tests using a Bonferroni correction. We modeled year as a random variable, which allowed us to pool the data across both years to obtain a single predicted slope for 24 points without ignoring the covariance among points within the same year. The slope was then tested to determine if it differed significantly from zero.

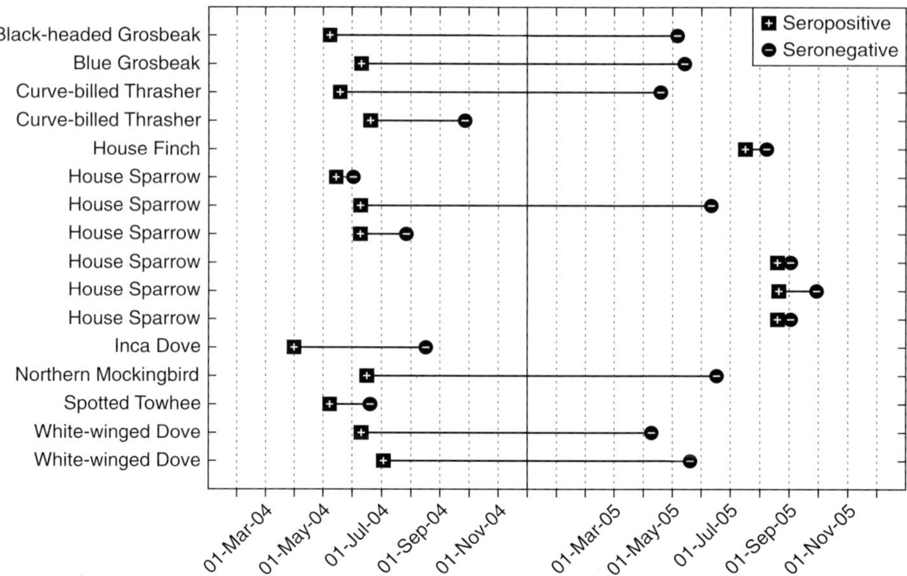

Figure 1.2. Dates of capture and recapture for the 16 individual birds that reverted from seropositive on initial capture to seronegative on a subsequent capture.

RESULTS

We obtained 1,977 samples in 2004 and 2,362 in 2005. The samples included 111 species from 32 families (Appendix 1.1). In 2004, the five most common species sampled included the House Sparrow (*Passer domesticus*, 23.5%), House Finch (*Carpodacus mexicanus*, 9.1%), White-winged Dove (*Zenaida asiatica*, 6.4%), Red-winged Blackbird (*Agelaius phoeniceus*, 4.3%), and Bullock's Oriole (*Icterus bullockii*, 4.1%). In 2005, the five most common species sampled included the House Sparrow (22.1%), House Finch (8.7%), Northern Mockingbird (*Mimus polyglottos*, 8.2%), Wilson's Warbler (*Wilsonia pusilla*, 6.3%), and Bullock's Oriole (3.9%). Highest levels of seroprevalence for individual species (combined across years where $n \geq 20$ samples) occurred in the Summer Tanager (*Piranga rubra*, 39%, $n = 28$), followed by American Robin (*Turdus migratorias*, 33%, $n = 64$), House Finch (17%, $n = 385$), Mourning Dove (*Zenaida macroura*, 14%, $n = 107$), and Pyrrhuloxia (*Cardinalis sinuatus*, 14%, $n = 37$). We detected 16 seroreversions, where birds were seropositive on the initial capture but seronegative on a subsequent capture (Fig. 1.2).

We performed a logistic regression to examine the main effects on seroprevalence in the avian community. We detected no significant differences between years (2004: 9.3%, $n = 1,977$; 2005: 5.8%, $n = 2,362$; $\chi^2 = 3.23$, df = 1, $P = 0.07$). Additionally, we found no differences between sexes ($\chi^2 = 0.04$, df = 1, $P = 0.84$), but differences in month, habitat, age, and residency status were significant ($\chi^2 = 33.72$, df = 5, $P < 0.0001$; $\chi^2 = 20.96$, df = 3, $P = 0.0001$; $\chi^2 = 8.54$, df = 1, $P = 0.0035$; $\chi^2 = 64.88$, df = 3, $P > 0.0004$, respectively).

We performed logistic regression to test for interactions among our main effects. We detected significant interactions for habitat–age–year ($\chi^2 = 34.93$, df = 4, $P < 0.0001$) and habitat–year ($\chi^2 = 15.56$, df = 3, $P = 0.001$), but not for habitat–age ($\chi^2 = 2.93$, df = 3, $P = 0.402$; Fig. 1.3). AHY birds had much higher seroprevalence than HY birds in all habitats except for desert in 2004. However, a decline in seroprevalence for AHY occurred in agricultural and riparian habitats from 2004 to 2005, and seroprevalence for HY birds increased in urban habitats from 2004 to 2005. In 2005, we did not detect differences between age groups across habitats.

We detected no significant interactions for month–age–year ($\chi^2 = 5.35$, df = 6, $P = 0.50$), but we did find significant effects for month–age ($\chi^2 = 26.5$, df = 5, $P < 0.0001$) and month–year ($\chi^2 = 52.62$, df = 5, $P < 0.0001$). In 2004, seroprevalence was highest for AHY birds early in the season (~15% ± 2.5), but then it declined through early fall, when seroprevalence was similar to HY

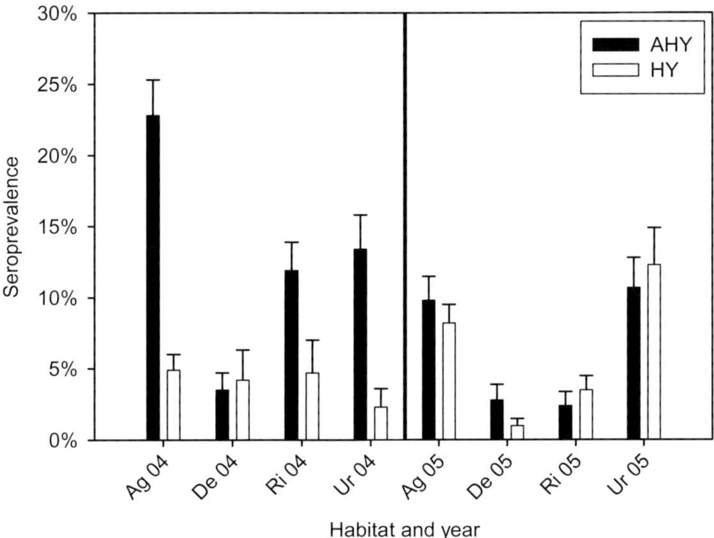

Figure 1.3. Comparison of seroprevalence (±SE) among habitats for AHY and HY birds in 2004 and 2005. In 2004, seroprevalence was highest in agricultural habitats, followed by riparian and urban habitats, and lowest in the desert habitat for AHY birds. Least square means found no differences between HY birds in 2004. In 2005, seroprevalence declined in the agricultural and riparian habitats for AHY birds, but seroprevalence for HY birds in 2005 increased in the agricultural and urban habitats. Ag = agriculture, De = desert, Ri = riparian, Ur = urban.

birds (~5% ± 2.2). In 2005, seroprevalence in both AHY and HY birds was low in spring (~6% ± 1.7) but then increased late in the summer (~10% ± 2.2).

We detected a significant sex–age–year interaction ($\chi^2 = 8.33$, df = 2, $P = 0.0155$, $n = 1,823$) as well as a significant sex–year interaction ($\chi^2 = 5.96$, df = 1, $P = 0.0147$), but the sex–age interaction was not significant ($\chi^2 = 0.13$, df = 1, $P = 0.716$). In 2004, AHY birds of both sexes had higher seroprevalence than HY birds, but there was no difference between sexes across age classes in 2005. Additionally, seroprevalence for adult males declined from 2004 to 2005 (12.8% ± 2.2 to 8.1% ± 1.1).

We recognized four categories of resident status: breeding, migrant, permanent resident, and winter resident. Migrants comprised about 11% (482 of 4,339) of all birds captured. Migrants included 25 species, with the five most common species being Wilson's Warbler ($n = 222$), Yellow Warbler (*Dendroica petechia*, $n = 70$), MacGillivray's Warbler (*Oporornis tolmiei*, $n = 54$), Orange-crowned Warbler (*Vermivora celata*, $n = 36$), and Virginia's Warbler (*V. virginiae*, $n = 24$). Abundance of wintering birds was similar, making up 10.5% (454 of 4,339) of our captures. Despite their common occurrence in our samples, there were only

three positive individuals each for the wintering and migrant categories over both years.

Tests of interaction effects for residency status only included resident and breeding birds because there were only three positive individuals each for the migrant and wintering bird categories. We detected differences across all levels of the interactions (status–year–age: $\chi^2 = 20.43$, df = 2, $P < 0.0001$; status–age: $\chi^2 = 15.85$, df = 1, $P < 0.001$; status–year: $\chi^2 = 7.41$, df = 1, $P = 0.007$; Fig. 1.4). In 2004, seroprevalence was highest for AHY resident birds compared to HY residents, but not different for either AHY or HY breeding birds in the same year. Seroprevalence for AHY residents and breeding birds declined from high levels in 2004 to much lower levels in 2005. Declines resulted in similar levels of seroprevalence in 2005 across age classes for both resident and breeding birds. In 2005, we failed to detect differences between age classes for both breeding and resident birds.

The logistic model indicated an age–year interaction ($\chi^2 = 13.44$, df = 1, $P = 0.0002$, $n = 4,070$). In 2004, AHY birds had higher seroprevalence than HY birds (13.2%, $n = 1,030$ vs. 4.3%, $n = 697$, respectively; $\chi^2 = 41.77$, df = 1, $P < 0.0001$), but age differences were not apparent in 2005 (6.7%,

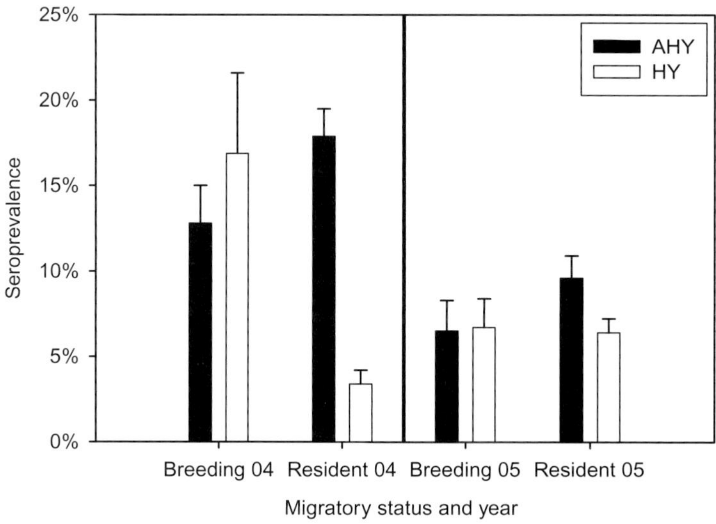

Figure 1.4. Comparisons of seroprevalence ($\pm$SE) among categories of migratory status by age for 2004 and 2005. Overall, seroprevalence dropped from 2004 to 2005 for both breeding and resident birds in all age classes except HY residents. Seroprevalence between AHY and HY breeding and resident birds in 2005 were not different from one another. Migrants and wintering birds were excluded from this analysis because only three seropositive individuals were in each category over both years.

$n = 1,003$ vs. 5.3%, $n = 1,340$, respectively; $\chi^2 = 1.96$, df = 1, $P = 0.1617$).

We selected four common bird species (Bullock's Oriole, Mourning Dove, Red-winged Blackbird, and House Finch) to illustrate seasonal patterns in abundance of age class in relation to rates of seroprevalence (Fig. 1.5). In both years, the proportion of AHY birds captured declined seasonally as the HY birds entered local populations. Seroprevalence in AHY birds tended to peak about a month earlier than in HY birds.

We compared measures of species diversity for avian communities at the 12 study sites with levels of seroprevalence for WNV. First, we modeled year as a random variable using a mixed linear model to account for covariance between years. We found that the estimate of covariance between years was much smaller (1–3 orders of magnitude) than the estimate of the residual variance. Thus, between-year variation had little effect on the model, allowing us to combine years in our analysis, providing 24 data points (12 study sites × 2 yr). We found a negative relationship between species diversity (H′) and seroprevalence ($F = 8.82$, df = 1, 21, $P = 0.0072$; Fig. 1.6a). Species richness showed a similar negative relationship with seroprevalence ($F = 13.62$, df = 1, 22, $P = 0.0013$; Fig. 1.6b). However, comparisons with the evenness measure (J′)

failed to show a relationship with seroprevalence ($F = 1.24$, df = 1, 21, $P = 0.2776$; Fig. 1.6c).

DISCUSSION

Variation Among Habitats

Comparisons across habitats indicated differences in the composition of avian communities as well as in rates of seroprevalence for individual bird species. The agricultural habitat had highest levels of seroprevalence overall, followed by the urban habitat. The riparian habitat had high seroprevalence in 2004 and low in 2005, and desert habitat was lowest overall (Fig. 1.3). Habitat-specific variation may result from many factors, including interactions between habitat availability for hosts and vectors, as well as characteristics of the avian and vector communities. Agricultural areas provide some of the best habitats in the region for both avian hosts and mosquito vectors. Agricultural areas supported the most seropositive hosts, including the Summer Tanager, House Sparrow, House Finch, European Starling (*Sturnus vulgaris*), and Mourning Dove. In addition, extensive irrigation systems in these areas provide an abundance of breeding habitats for mosquitoes throughout the long growing

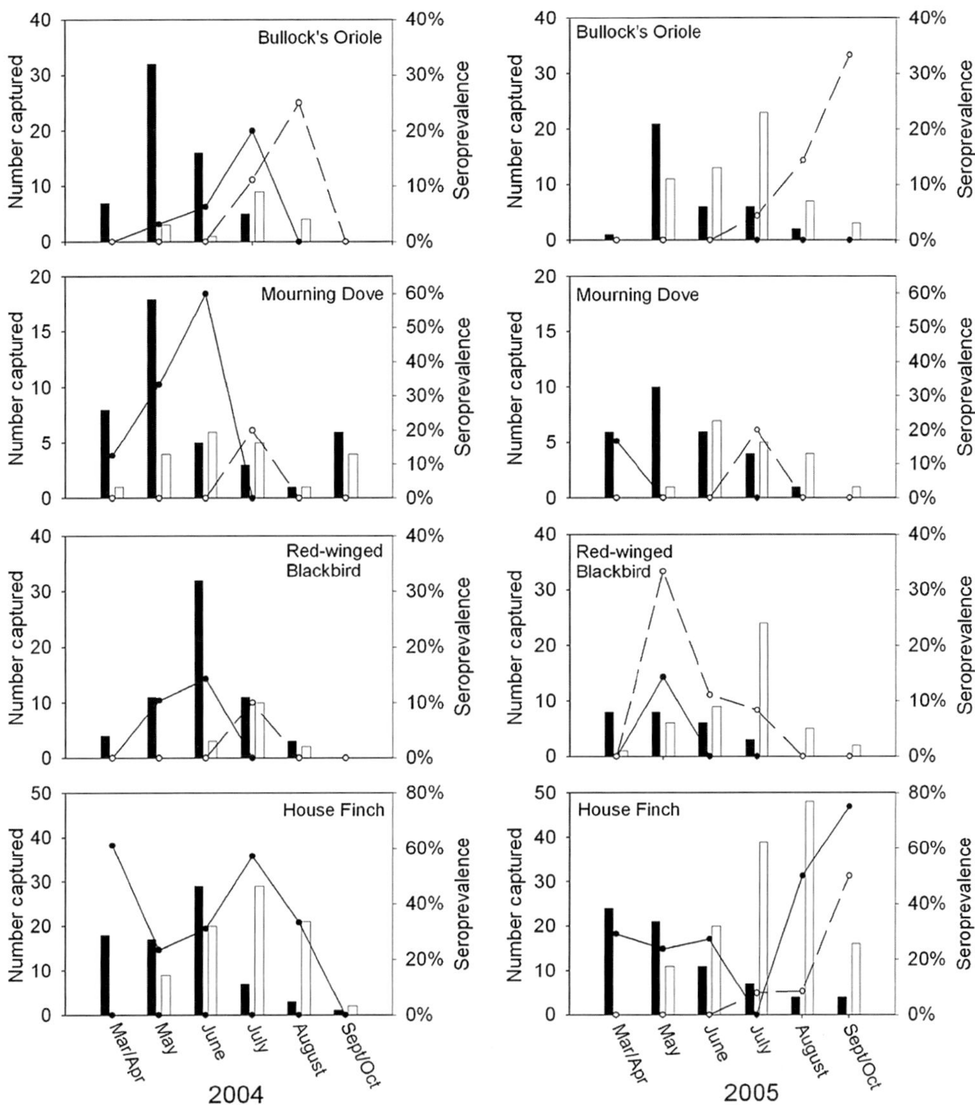

Figure 1.5. Examples of four common species comparing number of birds captured and seroprevalence for AHY and HY birds. Seroprevalence in AHY birds tended to peak in early summer, about a month before the peak in seroprevalence for HY birds. Solid bars are number of AHY birds captured; white bars are number of HY birds captured; solid line is seroprevalence of AHY birds; dashed line is seroprevalence HY birds. Note: Y-axis varies among species.

season. Seasonal overlap between birds and mosquitoes provides opportunity for frequent contact between avian hosts and mosquito vectors and likely contributes to the high levels of seroprevalence we found in agricultural areas.

We also found high levels of seroprevalence in urban habitats, as has been reported in other studies of WNV (Gruwell et al. 2000, Ringia et al. 2004, Reisen et al. 2005, Beveroth et al. 2006). Urban habitats generally have the densest human populations, support an abundance

of highly competent host species, and produce large numbers of the principal mosquito vectors. Abundant host species include the House Sparrow, American Robin, House Finch, Mourning Dove, and Northern Mockingbird. Anthropogenic water sources such as irrigation for ornamental plants, standing water associated with refuse, and water catchments and watering devices for pets provide an abundance of breeding habitats for mosquitoes. Many human activities in urban settings provide a quality environment for hosts and

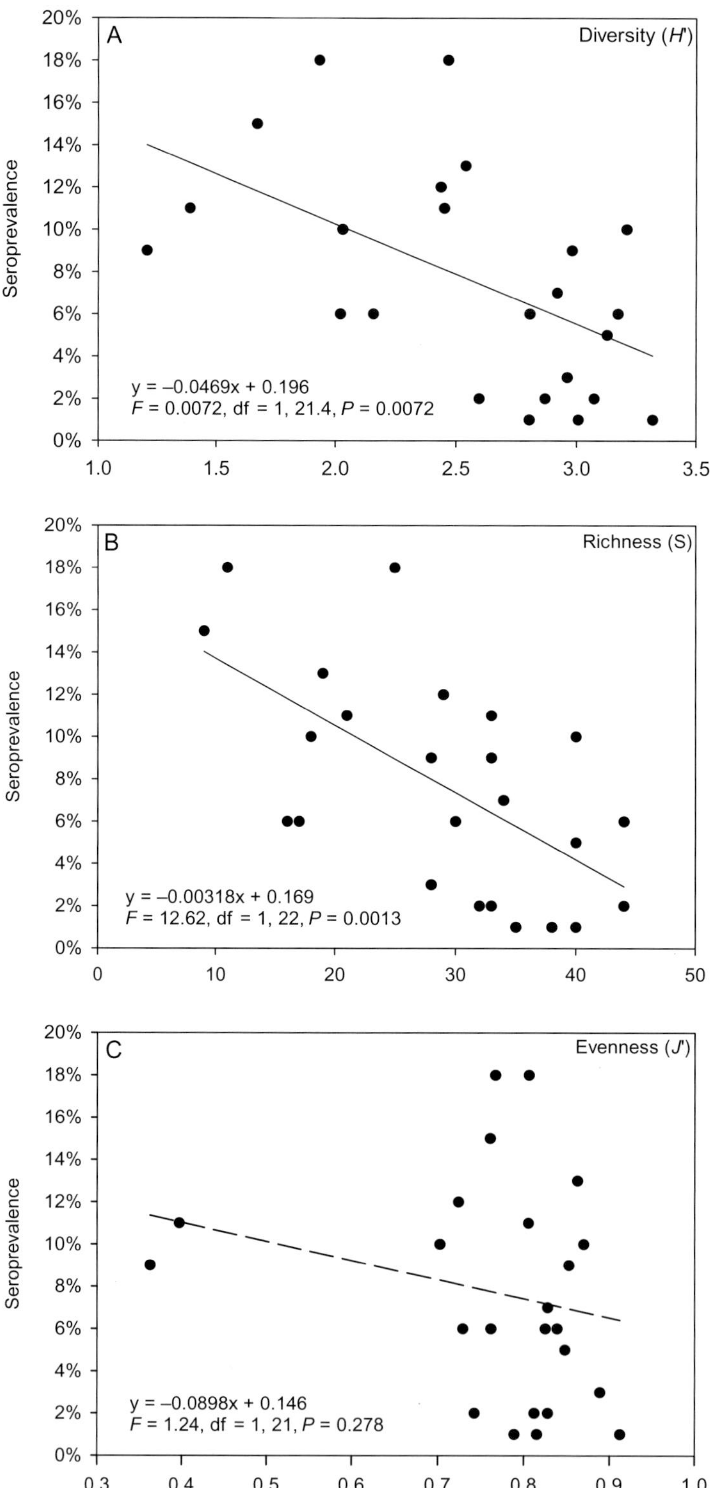

Figure 1.6. Relationships between measures of diversity of avian host communities in four habitat types during 2004 and 2005. Diversity (a) and richness (b) both have a negative relationship with seroprevalence. Analysis of evenness (c) and seroprevalence did not reveal a relationship, but a similar pattern was suggested, as denoted by the dashed line.

vectors, contributing to consistently high levels of seroprevalence in these areas.

Riparian habitat provided interesting results because of a large decrease in seroprevalence for AHY birds from 2004 (11.9%) to 2005 (2.4%). In a parallel study of mosquito vectors conducted at the same sites and in the same seasons, overall mosquito captures were much greater but had fewer mosquito pools positive for WNV in 2005 as compared to 2004 (Caccamise et al. 2006, Pitzer et al. 2009). In 2005, the lower proportion of positive mosquito pools could have contributed to the lower seroprevalence for avian hosts in the riparian habitat.

Desert sites had the lowest seroprevalence while supporting the highest levels of avian community richness. Although we chose sites in the desert that could provide aquatic habitat when summer rains occurred, availability of breeding sites for mosquitoes varied through the study. Sites often dried quickly after summer rains, making mosquito reproduction highly variable. Despite limited opportunities for virus transmission and generally low seroprevalence, some individual species in desert habitats had relatively high rates of seroprevalence, including the Pyrrhuloxia, Northern Mockingbird, and House Finch.

Patterns in Seroprevalence and Life History Traits

We found considerable variation in seroprevalence among the avian hosts we examined. American Robin and Summer Tanager had the highest average seroprevalence in 2005 (Appendix 1.1). American Robins are known to have only moderate seroprevalence (Beveroth et al. 2006) even though they are commonly fed on by mosquitoes and can act as superspreaders of WNV (Kilpatrick et al. 2006a, 2006b). Seroprevalence for Summer Tanagers in our study was much greater than infection rates found in Georgia (Gibbs et al. 2006). Although involvement of these two species in WNV cycling depends on levels of viremia and feeding preferences of bridge vectors, high seroprevalence of WNV suggests considerable potential for involvement in the virus cycle.

It is unclear why differences in seroprevalence occurred between sexes; this could have been an artifact of short-term sampling in our two-year study. On the other hand, Marshall et al. (2006) found a sex difference for Northern Cardinals, with females having higher WNV seroprevalence than males in another two-year study. The authors attributed the difference to behavioral differences between the sexes, because females incubate the eggs and therefore have a higher probability of being bitten by mosquitoes. Still, other studies comparing seroprevalence between sexes for WNV, SLE, EEE have not shown any sex differences (Crans et al. 1994, Gruwell et al. 2000, Reisen et al. 2001, Ringia et al. 2004, Beveroth et al. 2006).

We found little evidence supporting a role for neotropical migrants in initiating virus transmission along migratory routes into New Mexico. We detected only three seropositive individuals out of 482 migrants captured and tested. Dupuis et al. (2003) captured four seropositive migrants in Jamaica and Mexico, but failed to detect any viremic migrants. Crans et al. (1994), studying EEE in New Jersey, also suggested that migrants may not be important in the transmission of EEE. Infrequent detection of seropositive migrants may be partly due to difficulties in mounting an immune response while migrating, coupled with the resulting high mortality rates among infected birds (Male 2003). Other studies have proposed that mosquitoes transported along highways (Reisen et al. 2004a) and HY bird dispersal (Beveroth et al. 2006) are potentially more important factors in WNV spread.

Several past studies have shown that migrants can potentially transport viruses long distances. Fledgling White Storks (*Ciconia ciconia*) migrating from Europe were positive for WNV upon arrival to southern Israel (Malkinson et al. 2002). Along the eastern seaboard of the U.S., Lord and Calisher (1970) isolated EEE and WEE from migrants during fall migration. Although the number of individual migrants testing positive for virus was small, extrapolating the number of cases to the whole neotropical migratory population would result in large numbers of migrants transporting virus to wintering grounds (Lord and Calisher 1970). Peterson et al. (2003) showed through modeling that migratory birds are critical in the transport of WNV. Although we did not find much support for neotropical migrants initiating virus transmission in southern New Mexico, we found that breeding birds, which can be considered as migrants arriving primarily from Mexico and Central America, did have relatively high rates of seroprevalence.

Highest rates of seroprevalence occurred in resident and breeding birds. Our observations are similar to results from other studies of WNV and various other arboviruses (Work et al. 1955; Crans et al. 1994, Reisen et al. 2000, 2004b; Ringia et al. 2004; Beveroth et al. 2006). Opportunity for exposure to WNV is high for resident and breeding birds because these species are present in southern New Mexico throughout peak mosquito season (Caccamise et al. 2006, Pitzer et al. 2009). In addition, these species breed in areas where high numbers of vectors are present (e.g., *Culex pipiens*, *C. quinquifasciatus*; Caccamise et al. 2006, Pitzer et al. 2009), exposing susceptible HY birds to infection.

Hatch-year birds have the potential to prolong the virus season because young represent naïve, susceptible hosts as they hatch and recruit into the population (Hamer et al. 2008). Entrance of young into the avian community during the mosquito season provides feeding opportunities for mosquitoes and could drive the virus cycle to much higher levels in late summer (Reisen et al. 2000). Highest levels of seroprevalence for HY birds generally lagged about a month behind AHY birds (Fig. 1.5). The same seasonal trend was also present in four species examined for WEE in the Coachella Valley of California (Reisen et al. 2000).

Seroprevalence in AHY birds declined from 2004 to 2005, although seroprevalence for HY birds was similar in both years. Once infected, birds generally maintain antibodies to WNV from year to year, so the pool of AHY seropositive birds includes those infected in previous seasons and the current year (Beveroth et al. 2006). Circulating antibodies can produce an additive effect that would tend to inflate seroprevalence over time (Gruwell et al. 2000, Beveroth et al. 2006). However, WNV cycling has been shown to decline over several years after entering a newly susceptible population. In Illinois, Beveroth et al. (2006) detected a decrease in prevalence by the third year following initial establishment of the virus. Reisen et al. (1997) detected fewer seroconversions in sentinel chickens with SLE and WEE from 1991 to 1992 in the Imperial Valley of California, despite capturing >2 times the number of female mosquitoes per trap night in 1992. The first cases of WNV in humans were detected in New Mexico in 2003, so our study in 2005 likely represented the third year following introduction.

Declines in seroprevalence of AHY birds between 2004 and 2005 may represent a pattern similar to that found by Beveroth et al. (2006) in the third year following introduction. Decreasing incidence has been shown in other geographic areas for WNV with humans, birds, and horses (Peterson et al. 2004, CDC 2009) and for SLE in avian communities in California (Gruwell et al. 2000). Declines in seroprevalence in our bird population can also be attributed to reinfection of previously seropositive birds. Reinfection of hosts can result in brief immune responses, lowering the detectability of antibodies (Reisen et al. 2001, 2003). This immune response, coupled with antibody elimination, can result in false negatives. We detected 16 individuals positive at initial capture but negative when recaptured (Fig. 1.2). Reversals in seroprevalance have also been detected in House Finches (Gruwell et al. 2000) and sentinel chickens (HBV pers. obs.; L. Reed pers. comm.). In many cases, reversals can occur through the gradual decline in antibody titers, but other possibilities could include a reduction in antibody production in older infections, or when infections with low virus titers fail to elicit a full antibody response (Reisen et al. 2003).

Avian Diversity and WNV Seroprevalence

We found an inverse relationship between seroprevalence and both diversity and richness (Fig. 1.6a–b) over the 12 avian communities we studied. Desert sites had the highest diversity of birds and the lowest seroprevalence. High diversity resulted in part from large numbers of migrant and wintering resident species captured in desert habitats. In contrast, agricultural and urban sites had low diversity and high seroprevalence. Human-modified habitats had greater numbers of local and exotic species (e.g., House Sparrows, House Finches; Emlen 1974, Beissinger and Osborne 1982) that tend to be competent hosts for WNV (Komar et al. 2003). Riparian sites were somewhat unusual, because in both years diversity was high but seroprevalence was greater in 2004 than 2005. While avian community characteristics may influence rates of seroprevalence, other factors likely play a role as well. For example, higher abundances of competent host species in urban and agricultural habitats may be the key factor influencing seroprevalence rather than avian diversity. In addition, differences in the vector communities among study sites likely also contributed to the differences in seroprevalence we found, but the magnitude of these contributions remains unknown.

We did not detect a relationship between evenness and seroprevalence, though a trend of decreasing seroprevalence with increasing evenness was suggested (Fig. 1.6c). A two-species model for Lyme disease by Schmidt et al. (1999) detected lower proportions of feeding ticks in small mammal communities with increasing evenness when they held richness constant. Ostfeld and Keesing (2000) suggest that evenness is the most appropriate parameter in assessing disease risk and that evenness, like richness, plays an important role in disease risk. Failure to detect a relationship in our results may be related to the small number of communities we were able to sample ($n = 12$). With more intensive sampling, a relationship might be detected. The predicted relationship is that as bird diversity increases, the rate of exposure of bird species to infected mosquitoes decreases, resulting in a lower seroprevalence.

Two recent WNV studies have found contrasting results in testing the hypothesis for a relationship between diversity and disease risk. Allan et al. (2009) found increasing human incidence and prevalence of mosquito infection with decreasing bird diversity as well as increasing reservoir competency of the bird community. At the local level (St. Louis, MO; Allan et al. 2009), human-population density and community-competence index of the birds provided the greatest predictive power for WNV prevalence in mosquitoes. At the national level, human per capita incidence was positively associated with human density and community competence index, and negatively associated with bird diversity. Conversely, Loss et al. (2009) did not detect a net effect of increasing species richness on a reduction of WNV transmission in the greater Chicago, IL, metropolitan area. The authors found that avian seroprevalence was high overall and that high seroprevalence occurs across host species. Loss et al. suggested that other intrinsic and extrinsic factors such as variation in mosquito host preference, precipitation, temperature, and reservoir competence should be taken into account as important factors driving interannual WNV variation rather than the host diversity.

Our study indicates that host community composition may be related to disease risk within the community. A previous study by Ezenwa et al. (2006) also found support for the dilution hypothesis. However, they showed that species richness for non-passerines showed a negative correlation with human and mosquito infection rates of WNV, whereas the majority of the birds in our studies were passerines. While it appears that support continues for the dilution hypothesis, a need remains to examine other factors that might drive WNV cycles. One such factor is the dynamics of vector populations in the overall analyses of WNV prevalence in host species. Only a combined approach that includes vectors and hosts can provide insight into mechanics of the transmission cycle, and how these interact with host and vector populations to affect disease risk.

ACKNOWLEDGMENTS

We are particularly grateful to M. Jozan, who provided essential guidance in laboratory methods and contributed intellectually and inspirationally throughout. R. Hall generously provided antigen and monoclonal antibodies for the ELISA tests. We thank K. Hanley and W. Whitford for the many contributions they made to all aspects of this project, as well as R. Ostfeld for comments and suggestions. Appreciation extends to L. Reed and R. Puelle for providing WNV positive and negative sera. We wish to acknowledge the valuable contributions provided by anonymous reviewers, who helped substantially in improving the manuscript. We also thank the many individuals who contributed to all aspects of the field work as well as the land owners, including USDA Agriculture Research Service Jornada Experimental Range and Chihuahuan Desert Rangeland Research Center, for providing access to the study sites. The Jornada Experimental Range is administered by the USDA-ARS and is a Long Term Ecological Research site funded by the National Science Foundation. Funding was provided by New Mexico State Department of Health, T&E Inc., and the New Mexico Agricultural Experiment Station.

LITERATURE CITED

Allan, B., R. Langerhans, W. Ryberg, W. Landesman, N. Griffin, R. Katz, B. Oberle, M. Schutzenhofer, K. Smyth, A. De St. Maurice, L. Clark, K. Crooks, D. Hernandez, R. Mcclean, R. Ostfeld, and J. Chase. 2009. Ecological correlates of risk and incidence of West Nile virus in the United States. Oecologia 158:699–708.

Beissinger, S. R., and D. R. Osborne. 1982. Effects of urbanization on avian community organization. Condor 84:75–83.

Bernard, K. A., J. G. Maffei, S. A. Jones, E. B. Kauffman, G. D. Ebel, A. P. Dupuis, K. A. Ngo, D. C. Nicholas, D. M. Young, P. Y. Shi, V. L. Kulasekera, M. Eidson, D. J. White, W. B. Stone, and L. D. Kramer. 2001. West Nile virus infection in birds and mosquitoes, New York State, 2000. Emerging Infectious Diseases 7:679–685.

Beveroth, T. A., M. P. Ward, R. L. Lampman, A. M. Ringia, and R. J. Novak. 2006. Changes in seroprevalence of West Nile virus across Illinois in free-ranging birds from 2001 through 2004. American Journal of Tropical Medicine and Hygiene 74: 174–179.

Caccamise, D. F, R. Creamer, R. Byford, H. B. Vuong, and J. B. Pitzer. 2006. West Nile virus: patterns in the establishment and maintenance of an exotic pathogen in an arid landscape. Final report to New Mexico Department of Health, Santa Fe, NM.

Centers for Disease Control. 2009. West Nile virus homepage. <www.cdc.gov/ncidod/dvbid/westnile/index.htm>.

Crans, W. J., D. F. Caccamise, and J. R. McNelly. 1994. Eastern equine encephalomyelitis virus in relation to the avian community of a coastal cedar swamp. Journal of Medical Entomology 31:711–728.

Dick-Peddie, W. A. 1999. New Mexico vegetation: past, present, and future. University of New Mexico Press, Albuquerque, NM.

Dupuis, A. P., P. P. Marra, and L. D. Kramer. 2003. Serologic evidence of West Nile virus transmission, Jamaica, West Indies. Emerging Infectious Diseases 9:860–863.

Eidson, M., N. Komar, F.Sorhage, R. Nelson, T. Talbot, F. Mostashari, and R. McLean. 2001a. Crow deaths as a sentinel surveillance system for West Nile virus in the northeastern United States, 1999. Emerging Infectious Diseases 7:615–620.

Eidson, M., L. Kramer, W. Stone, Y. Hagiwara, and K. Schmit. 2001b. Dead bird surveillance as an early warning system for West Nile virus. Emerging Infectious Diseases 7:631–635.

Emlen, J. T. 1974. An urban bird community in Tucson, Arizona: derivation, structure, regulation. Condor 76:184–197.

Ezenwa, V. O., M. S. Godsey, R. J. King, and S. C. Guptill. 2006. Avian diversity and West Nile virus: testing associations between biodiversity and infectious disease risk. Proceedings of the Royal Society B—Biological Sciences 273:109–117.

Gibbs, S. E, J. A. Allison, M. J. Yabsley, D. G. Mead, B. R. Wilcox, and D. E. Stallknecht. 2006. West Nile virus antibodies in avian species of Georgia, USA: 2000–2004. Vector-Borne and Zoonotic Diseases 6:57–72.

Goddard, L. B., A. E. Roth, W. K. Reisen, and T. W. Scott. 2002. Vector competence of California mosquitoes for West Nile virus. Emerging Infectious Diseases 8:1385–1391.

Gruwell, J. A., C. L. Fogarty, S. G. Bennett, G. L. Challet, K. S. Vanderpool, M. Jozan, and J. P. Webb. 2000. Role of peridomestic birds in the transmission of St. Louis encephalitis virus in southern California. Journal of Wildlife Diseases 36:13–34.

Hamer, G. L., E. D. Walker, J. D. Brawn, S. R. Loss, M. O. Ruiz, T. L. Goldberg, A. M. Schotthoefer, W. M. Brown, E. Wheeler, and U. D. Kitron. 2008. Rapid amplification of West Nile virus: the role of hatch-year birds. Vector-Borne and Zoonotic Diseases 8:57–67.

Harris, M. C., and J. M. Sleeman. 2007. Morbidity and mortality of Bald Eagles (*Haliaeetus leucocephalus*) and Peregrine Falcons (*Falco peregrinus*) admitted to the Wildlife Center of Virginia, 1993–2003. Journal of Zoo and Wildlife Medicine 38:62–66.

Jozan, M., R. Evans, R. McClean, R. Hall, B. Tangredi, L. Reed, and J. Scott. 2003. Detection of West Nile virus infection in birds in the United States by blocking ELISA and immunohistochemistry. Vector-Borne and Zoonotic Diseases 3:99–110.

Kilpatrick, A. M., P. Daszak, M. J. Jones, P. P. Marra, and L. D. Kramer. 2006a. Host heterogeneity dominates West Nile virus transmission. Proceedings of the Royal Society B—Biological Sciences 273:2327–2333.

Kilpatrick, A. M., L. D. Kramer, M. J. Jones, P. P. Marra, and P. Daszak. 2006b. West Nile virus epidemics in North America are driven by shifts in mosquito feeding behavior. PLoS Biology 4:606–610.

Komar, N., S. Langevin, S. Hinten, N. Nemeth, E. Edwards, D. Hettler, B. Davis, R. Bowen, and M. Bunning. 2003. Experimental infection of North American birds with the New York 1999 strain of West Nile virus. Emerging Infectious Diseases 9:311–322.

LoGuidice, K., R. S. Ostfeld, K. A. Schmidt, and F. Keesing. 2003. The ecology of infectious disease: effects of host diversity and community composition on Lyme disease risk. Proceedings of the National Academy of Sciences of the United States of America 100:567–571.

Lord, R. D., and C. H. Calisher. 1970. Further evidence of southward transport of arboviruses by migratory birds. American Journal of Epidemiology 92:73–78.

Loss, S., G. Hamer, E. Walker, M. Ruiz, T. Goldberg, U. Kitron, and J. Brawn. 2009. Avian host community structure and prevalence of West Nile virus in Chicago, Illinois. Oecologia 159:415–424.

Magurran, A. E. 1998. Ecological diversity and its measurement. Princeton University Press, Princeton, NJ.

Male, T. 2003. Potential impact of West Nile virus on American avifaunas. Conservation Biology 17:928–930.

Malkinson, M., C. Banet, Y. Weisman, S. Pokamunski, R. King, M. T. Drouet, and V. Deubel. 2002. Introduction of West Nile virus in the Middle East by migrating White Storks. Emerging Infectious Diseases 8:392–397.

Marshall, J. S., D. A. Zuwerink, R. A. Restifo, and T. C. Grubb. 2006. West Nile virus in the permanent-resident bird community of a fragmented Ohio landscape. Ornithological Monographs 60:79–85.

McIntosh, B. M., and P. G. Jupp. 1982. Ecological studies on West Nile virus in southern Africa. Arbovirus Research in Australia: Proceedings of 3rd Symposium, February 15–17, 1981.

Mills, J. N. 2006. Biodiversity loss and emerging infectious disease: an example from the rodent-borne hemorrhagic fevers. Biodiversity 7:9–17.

Ostfeld, R., and F. Keesing. 2000. The function of biodiversity in the ecology of vector-borne zoonotic diseases. Canadian Journal of Zoology 78:2061–2078.

Peterson, A. T., N. Komar, O. Komar, A. Navarro-Siguenza, M. B. Robbins, and E. Martinez-Meyer. 2004. West Nile virus in the New World: potential impacts on bird species. Bird Conservation International 14:215–232.

Peterson, A. T., D. A. Vieglais, and J. K. Andreasen. 2003. Migratory birds modeled as critical transport agents for West Nile virus in North America. Vector-Borne and Zoonotic Diseases 3:27–37.

Pitzer, J. B., R. L. Byford, H. B. Vuong, R. L. Steiner, R. J. Creamer, and D. F. Caccamise. 2009. Potential vectors of West Nile virus in a semiarid environment: Doña Ana County, New Mexico. Journal of Medical Entomology 46:1474–1482.

Pyle, P. 1997. Identification guide to North American birds, Part I: Columbidae to Ploceidae, Slate Creek Press, Bolinas, CA.

Ralph, C. J., G. R. Geupel, P. Pyle, T. E. Martin, and D. F. DeSante. 1993. Handbook of field methods for monitoring landbirds. USDA Forest Service, General Technical Report PSW-GTR-144, USDA Forest Service, Pacific Southwest Research Station, Albany, CA.

Rappole, J. H., S. R. Derrickson, and Z. Hubalek. 2000. Migratory birds and spread of West Nile virus in the Western Hemisphere. Emerging Infectious Diseases 6:319–328.

Reed, L., and W. J. Crans. 1998. Geographical variation in avian migration cycles in an eastern equine encephalitis region. Pp. 29–34 *in* Proceedings of the Annual Meeting of the Northeastern Mosquito Control Association, December.

Reisen, W. K., H. D. Lothrop, S. B. Presser, J. L. Hardy, and E. W. Gordon. 1997. Landscape ecology of arboviruses in southeastern California: temporal and spatial patterns of enzootic activity in Imperial Valley, 1991–1994. Journal of Medical Entomology 34:179–188.

Reisen, W. K., J. O. Lundstrom, T. W. Scott, B. F. Eldridge, R. E. Chiles, R. Cusack, V. M. Martinez, H. D. Lothrop, B. Gutierrez, S. E. Wright, K. Boyce, and B. R. Hill. 2000. Patterns of avian seroprevalence to western equine encephalomyelitis and Saint Louis encephalitis viruses in California, USA. Journal of Medical Entomology 37:507–527.

Reisen, W. K., L. D. Kramer, R. E. Chiles, E. N. Green, and V. M. Martinez. 2001. Encephalitis virus persistence in California birds: preliminary studies with House Finches. Journal of Medical Entomology 38:393–399.

Reisen, W. K., R. E. Chiles, V. M. Martinez, Y. Fang, and E. N. Green. 2003. Experimental infection of California birds with western equine encephalomyelitis and St. Louis encephalitis viruses. Journal of Medical Entomology 40:968–982.

Reisen, W., H. Lothrop, R. Chiles, L. Woods, M. Madon, C. Cossen, S. Husted, V. Kramer, and J. Edman. 2004a. West Nile virus in California. Emerging Infectious Diseases 10:1369–1378.

Reisen, W. K., R. E. Chiles, V. M. Martinez, Y. Fang, and E. N. Green. 2004b. Encephalitis virus persistence in California birds: experimental infections in mourning doves (Zenaida macroura). Journal of Medical Entomology 41:462–466.

Reisen, W. K., Y. Fang, and V. M. Martinez. 2005. Avian host and mosquito (Diptera: Culicidae) vector competence determine the efficiency of West Nile and St. Louis encephalitis virus transmission. Journal of Medical Entomology 42:367–375.

Ringia, A. M., B. J. Blitvich, H. Y. Koo, M. Van De Wyngaerde, J. D. Brawn, and R. J. Novak. 2004. Antibody prevalence of West Nile virus in birds, Illinois, 2002. Emerging Infectious Diseases 10:1120–1124.

SAS Institute Inc. Cary, NC, USA

Schmidt, K. A., R. S. Ostfeld, and E. M. Schauber. 1999. Infestation of *Peromyscus leucopus* and *Tamias striatus* by *Ixodes scapularis* (Acari: Ixodidae) in relation to the abundance of hosts and parasites. Journal of Medical Entomology 36:749–757.

Sibley, D. A. 2000. The Sibley guide to birds. Chanticleer Press, New York, NY.

Taylor, R. M., T. H. Work, H. S. Hurlburt, and F. Rizk. 1956. A study of the ecology of West Nile virus in Egypt. American Journal of Tropical Medicine and Hygiene 5:579–620.

Thompson, B. C., D. A. Leal, and R. A. Meyer. 1994. Bird community composition and habitat importance in the Rio Grande system of New Mexico with emphasis on neotropical migrant birds. USDI Fish and Wildlife Service Report, Region 2, Division of Wildlife and Habitat Management, Albuquerque, NM.

Tsai, T. F., F. Popovici, C. Cernescu, G. L. Campbell, and N. I. Nedelcu. 1998. West Nile encephalitis epidemic in southeastern Romania. Lancet 352:767–771.

Western Regional Climate Center. 2009. New Mexico climate summaries. <www.wrcc.dri.edu/summary/climsmnm.html> (30 June 2009).

Wilcox, B. R., M. J. Yabsley, A. E. Ellis, D. E. Stallknecht, and S. E. Gibbs. 2007. West Nile virus antibody prevalence in American Crows (*Corvus brachyrhynchos*) and Fish Crows (*Corvus ossifragus*) in Georgia, USA. Avian Diseases 51:125–128.

Work, T.H., H.S. Hurlburt, and R.M. Taylor. 1955. Indigenous wild birds of the Nile Delta as potential West Nile virus circulating reservoirs. The American Journal of Tropical Medicine and Hygiene 4: 872–888.

All bird species captured and tested for WNV antibodies in southern New Mexico, 2004–2005.

Species	Status[a]	Habitat[b]	Number tested	Number positive	Seroprevalence
Gambel's Quail (*Callipepla gambelii*)	R	A,D,R	36	4	11%
Mourning Dove (*Zenaida macroura*)	R	A,D,R,U	107	15	14%
White-winged Dove (*Zenaida asiatica*)	R	A,D,R,U	204	6	3%
Inca Dove (*Columbina inca*)	R	U	55	1	2%
Rock Pigeon[c] (*Columbia livia*)	R	A	7		0%
Ladder-backed Woodpecker (*Picoides scalaris*)	R	A,D,R,U	17	3	18%
Western Wood-Pewee (*Contopus sordidulus*)	B	A,D,R	13		0%
Cordilleran Flycatcher (*Empidonax occidentalis*)	M	D,R,U	8		0%
Black Phoebe (*Sayornis nigricans*)	R	A,D,R,U	13	1	8%
Say's Phoebe (*Sayornis saya*)	R	A,D,R	6		0%
Ash-throated Flycatcher (*Myiarchus cinerascens*)	B	D,R	69		0%
Western Kingbird (*Tyrannus verticalis*)	B	A,D,R,U	49		0%
Loggerhead Shrike (*Lanius ludovicianus*)	R	A,D	17		0%
Warbling Vireo (*Vireo gilvus*)	M	D,R,U	18		0%
Northern Rough-winged Swallow (*Stelgidopteryx serripennis*)	B	D,R	15		0%
Cliff Swallow (*Petrochelidon pyrrhonota*)	B	D	6	2	33%
Barn Swallow (*Hirundo rustica*)	B	A,D,R,U	16	2	13%
Verdin (*Auriparus flaviceps*)	R	A,D,R,U	44	1	2%
Cactus Wren (*Campylorhynchus brunneicapillus*)	R	D,U	12		0%
House Wren (*Troglodytes aedon*)	R	A,D,R	5		0%
Marsh Wren (*Cistothorus palustris*)	W	R	7		0%
Ruby-crowned Kinglet (*Regulus calendula*)	W	A,D,R,U	20		0%
American Robin (*Turdus migratorius*)	B	A,R,U	64	21	33%
Northern Mockingbird (*Mimus polyglottos*)	R	A,D,R,U	274	10	4%
Curve-billed Thrasher (*Toxostoma curvirostre*)	R	A,D,U	57	7	12%
Crissal Thrasher (*Toxostoma crissale*)	R	A,D,R	11	3	27%
European Starling (*Sturnus vulgaris*)	R	A,U	46	2	4%

APPENDIX 1.1 (*continued*)

Species	Status[a]	Habitat[b]	Number tested	Number positive	Seroprevalence
Orange-crowned Warbler (*Vermivora celata*)	M	A,D,R	36		0%
Virginia's Warbler (*Vermivora virginiae*)	M	A,D,R	24	1	4%
Lucy's Warbler (*Vermivora luciae*)	M	D,R	11		0%
Yellow Warbler (*Dendroica petechia*)	M	A,D,R,U	70		0%
Yellow-rumped Warbler (*Dendroica coronata*)	W	A,D,R,U	36		0%
MacGillivray's Warbler (*Oporornis tolmiei*)	M	A,D,R,U	54	1	2%
Common Yellowthroat (*Geothlypis trichas*)	B	A,D,R,U	56	2	4%
Wilson's Warbler (*Wilsonia pusilla*)	M	A,D,R,U	222		0%
Yellow-breasted Chat (*Icteria virens*)	B	R,U	46	5	11%
Green-tailed Towhee (*Pipilo chlorurus*)	W	D,R,U	12		0%
Spotted Towhee (*Pipilo maculatus*)	R	D,R,U	10	1	10%
Cassin's Sparrow (*Aimophila cassinii*)	R	D	7		0%
Chipping Sparrow (*Spizella passerina*)	W	A,D,R,U	83	1	1%
Clay-colored Sparrow (*Spizella pallida*)	M	A,D,R	7		0%
Brewer's Sparrow (*Spizella breweri*)	W	A,D,R	78	1	1%
Vesper Sparrow (*Pooecetes gramineus*)	W	D,R	14		0%
Lark Sparrow (*Chondestes grammacus*)	R	A,D,R	52		0%
Black-throated Sparrow (*Amphispiza bilineata*)	R	D,R	24		0%
Lark Bunting (*Calamospiza melanocorys*)	W	A,D	10		0%
Song Sparrow (*Melospiza melodia*)	W	A,D,R	7		0%
Lincoln's Sparrow (*Melospiza lincolnii*)	W	A,D,R	29	1	3%
White-crowned Sparrow (*Zonotrichia leucophrys*)	W	A,D,R,U	135		0%
Dark-eyed Junco (*Junco hyemalis*)	W	A,D,R,U	16		0%
Summer Tanager (*Piranga rubra*)	B	A,R,U	28	11	39%
Western Tanager (*Piranga ludoviciana*)	B	A,D,R,U	16		0%
Pyrrhuloxia (*Cardinalis sinuatus*)	R	A,D,R,U	37	5	14%
Black-headed Grosbeak (*Pheucticus melanocephalus*)	B	A,D,R,U	63	7	11%

APPENDIX 1.1 (*continued*)

Species	Status[a]	Habitat[b]	Number tested	Number positive	Seroprevalence
Blue Grosbeak (*Passerina caerulea*)	B	A,D,R	73	6	8%
Lazuli Bunting (*Passerina amoena*)	M	D	5		0%
Painted Bunting (*Passerina ciris*)	B	A,R	8		0%
Red-winged Blackbird (*Agelaius phoeniceus*)	R	A,D,R	157	14	9%
Meadowlarks (*Sturnella* spp.)	R	D	11		0%
Great-tailed Grackle (*Quiscalus mexicanus*)	R	A,R,U	8		0%
Bronzed Cowbird (*Molothrus aeneus*)	B	A,U	11	3	27%
Brown-headed Cowbird (*Molothrus ater*)	R	A,D,R	44	1	2%
Hooded Oriole (*Icterus cucullatus*)	B	A,D	6		0%
Bullock's Oriole (*Icterus bullockii*)	B	A,D,R,U	174	8	5%
Scott's Oriole (*Icterus parisorum*)	B	D	5		0%
Unknown Oriole (*Icterus* sp.)	B	D	6		0%
House Finch (*Carpodacus mexicanus*)	R	A,D,R,U	385	66	17%
Lesser Goldfinch (*Spinus psaltria*)	R	A,U	7		0%
House Sparrow (*Passer domesticus*)	R	A,D,R,U	988	105	11%
Other species[d]	B,M,R,W	A,D,R,U	72	4	6%

[a] Migratory status of the birds included: B = breeding, M = migrant, R = resident, W = winter.

[b] Habitat where birds were captured included: A = agriculture, D = desert, R = riparian, U = urban.

[c] All Rock Pigeons sampled were captive birds.

[d] Species with fewer than five tested individuals (common name, scientific name, number tested, number positive): Least Bittern (*Ixobrychus exilis*, 2, 1); Green Heron (*Butorides virescens*, 2, 1); Sharp-shinned Hawk (*Accipiter striatus*, 1); Cooper's Hawk (*Accipiter cooperii*, 2, 1); Northern Bobwhite (*Colinus virginianus*, 4); Chukar (*Alectoris chukar*, 3); Ring-necked Pheasant (*Phasianus colchicus*, 1, 1); Killdeer (*Charadrius vociferous*, 2); Solitary Sandpiper (*Tringa solitaria*, 3); Eurasian Collared-Dove (*Streptopelia decaocta*, 2); Yellow-billed Cuckoo (*Coccyzus americanus*, 1); Greater Roadrunner (*Geococcyx californianus*, 1); Lesser Nighthawk (*Chordeiles acutipennis*, 1); Belted Kingfisher (*Ceryle alcyon*, 2); Northern Flicker (Red-shafted) (*Colaptes auratus*, 1); Willow Flycatcher (*Empidonax traillii*, 1); Dusky Flycatcher (*Empidonax oberholseri*, 4); Gray Flycatcher (*Empidonax wrightii*, 2); Bell's Vireo (*Vireo bellii*, 1); Western Scrub-Jay (*Aphelocoma californica*, 1); Bank Swallow (*Riparia riparia*, 1); Bewick's Wren (*Thryomanes bewickii*, 1); Rock Wren (*Salpinctes obsoletus*, 1); Black-tailed Gnatcatcher (*Polioptila melanura*, 3); unknown Gnatcatcher (*Polioptila* sp., 1); Swainson's Thrush (*Catharus ustulatus*, 1); Hermit Thrush (*Catharus guttatus*, 3); Gray Catbird (*Dumetella carolinensis*, 1); American Pipit (*Anthus rubescens*, 1); Phainopepla (*Phainopepla nitens*, 4); Black-throated Gray Warbler (*Dendroica nigrescens*, 1); Townsend's Warbler (*Dendroica townsendi*, 2); Black-and-white Warbler (*Mniotilta varia*, 1); American Redstart (*Setophaga ruticilla*, 1); Northern Waterthrush (*Seiurus noveboracensis*, 4); Mourning Warbler (*Oporonis philadelphia*, 1); Hooded Warbler (*Wilsonia citrina*, 1); Dickcissel (*Spiza americana*, 1); Savannah Sparrow (*Passerculus sandwichensis*, 2); White-throated Sparrow (*Zonotrichia albicollis*, 1); Orchard Oriole (*Icterus spurious*, 1); Pine Siskin (*Carduelis pinus*, 2).

CHAPTER TWO

The Trans-Atlantic Movement of the Spirochete *Borrelia garinii*

THE ROLE OF TICKS AND THEIR SEABIRD HOSTS

*Sabir Bin Muzaffar, Robert P. Smith, Jr., Ian L. Jones,
Jennifer Lavers, Eleanor H. Lacombe, Bruce K. Cahill,
Charles B. Lubelczyk, and Peter W. Rand*

Abstract. The spirochete *Borrelia garinii*, one of three genospecies of *B. burgdorferi* sensu lato (*B. burdorferi* s.l.) that can cause Lyme disease in humans, has recently been isolated from seabirds from a colony in Newfoundland, Canada. Previous records of *B. garinii* in seabirds suggest that it has been endemic in seabird colonies in the greater North Atlantic since at least the early 1990s. We determined the prevalence of *B. garinii* in different seabird hosts from colonies in the northwest Atlantic. We recorded *B. garinii* from Gannet Islands, Labrador, and Gull Island, Newfoundland, Canada, in Atlantic Puffins (*Fratercula arctica*), Herring Gulls (*Larus argentatus*), Common Murres (*Uria aalge*), and Razorbills (*Alca torda*). Prevalence of infections varied between years and within and among species. Ticks from Atlantic Puffins had a prevalence ranging from 10.3 to 36.4%, although the highest prevalence was noted in Herring Gulls (37.5%) in 2005. Earlier studies from the same localities failed to find evidence of *B. garinii*, suggesting a recent arrival of the spirochete into the northwest Atlantic. *B. garinii* is closely related to European strains of the spirochete, and its likely source is from areas of endemicity in the Bothnian Gulf and the northeast Atlantic seabird colonies where seabirds, songbirds, and two different tick species come in close proximity. Phylogenetic studies suggest a gradual movement of the European strains into seabird colonies in the northeast Atlantic with subsequent spread into the North and northwest Atlantic colonies. Atlantic Puffins seem to be suitable reservoirs, although other abundant species such as Common Murres and Thick-billed Murres (*Uria lomvia*) may be involved in *B. garinii* dynamics. Further work is urgently needed to help document the ecology and spread of this spirochete of importance to human health.

Key Words: Acari, *Borrelia burgdorferi*, *Borrelia garinii*, introduced rodents, *Ixodes uriae*, oceanic islands, seabirds, spread, ticks.

El Movimiento Tras-Atlántico de la Espiroqueta *Borrelia garinii*: El Papel de las Garrapatas y las Aves Marinas Como sus Hospederos

Resumen. La espiroqueta *Borrelia garinii*, una de las tres especies de *B. burgdorferi* sensu lato (*B. burgdorferi* s.l.) que causan la enfermedad de Lyme o borreliosis en humanos, ha sido recientemente aislada de aves marinas de una colonia en Newfoundland, Canada. Los registros previos de *B. garinii* en aves marinas sugieren que dicho parásito ha sido endémico en las colonias de aves marinas en el gran Atlántico Norte desde, al menos, principios de los 1990s. Determinamos

Muzaffar, S. B., R. P. Smith Jr., I. L. Jones, J. Lavers, E. H. Lacombe, B. K. Cahill, C. B. Lubelczyk, and P. W. Rand. 2012. Trans-Atlantic movement of the spirochete *Borrelia garinii*: the role of ticks and their seabird hosts. Pp. 23–30 *in* E. Paul (editor). Emerging avian disease. Studies in Avian Biology (vol. 42), University of California Press, Berkeley, CA.

la prevalencia de *B. garinii* en diferentes aves marinas en colonias del noroeste del Atlántico. Registramos *B. garinii*, en las Islas Gannet, Labrador y Gull, pertenecientes a Newfoundland, Canada, infectando frailecillos comunes (*Fratercula arctica*), gaviotas argénteas (*Larus argentatus*), araos comunes (*Uria aalge*) y alcas comunes (*Alca torda*). La prevalencia de infección varió entre años, entre especies y dentro de cada especie. Las garrapatas que parasitaron a los frailecillos comunes tuvieron una prevalencia de 10.3–36.4%, aunque la prevalencia más alta fue registrada en gaviotas argénteas (37.5%) en 2005. Estudios previos realizados en las mismas localidades no encontraron evidencia de *B. garinii*, lo que sugiere una llegada reciente de la espiroqueta en el Atlántico Noroeste. *B. garinii* esta cercanamente relacionada con cepas Europeas de la espiroqueta y su fuente de procedencia es posiblemente las áreas del Golfo de Bothnian y las colonias de aves marinas del Atlántico Noreste, en donde las aves marinas, las aves canoras y dos especies de garrapatas están muy cercanas unas a otras. Los estudios filogenéticos sugieren un movimiento gradual de las cepas Europeas hacia las colonias de aves marinas en el Atlántico Noreste, con una expansión subsecuente hacia las colonias del Atlántico Norte y Noroeste. Los frailecillos comunes parecen ser reservorios efectivos de la enfermedad, aunque otras especies abundantes, tales como el Arao común y el Arao de Brunnich o pico ancho (*Uria lomvia*), pueden estar involucradas en la dinámica de *B. garinii*. Se necesitan estudios urgentes que ayuden a documentar la ecología y la expansión de esta espiroqueta que tiene importancia para la salud humana.

Palabras Clave: Acari, aves marinas, *Borrelia burgdorferi*, *Borrelia garinii*, expansión, garrapatas, islas oceánicas, *Ixodes uriae*, roedores introducidos.

Borrelia burgdorferi sensu lato (s.l.) is a spirochete (Spirochaetes, Spirochaetaceae) that causes Lyme borreliosis in North America and Eurasia (Burgdorfer et al. 1982, Peisman and Gern 2004). At least 12 genospecies are recognized within *B. burgdorferi* s.l. The transmission cycle of *B. burgdorferi* s.l. involves ticks of the genus *Ixodes* (Acari: Ixodidae) and various mammalian or avian hosts. *Borrelia afzelii*, *B. garinii*, and *B. burgdorferi* sensu stricto are the only genotypes known to cause Lyme disease in humans, and the extent of the disease varies when caused by different strains within each genospecies (Kurtenbach et al. 2002, Lagal et al. 2003, Peisman and Gern 2004). *Borrelia garinii* strains usually circulate in cycles involving birds and rodents.

The dominant genospecies in North America is *B. burgdorferi* sensu stricto (s.s.), and although it primarily circulates in ticks and mammals (Rand et al. 2003, Peisman and Gern 2004), the spirochete is also abundant in a range of passerine songbirds (Anderson et al. 1986, Weisbrod and Johnson 1989, Magnarelli et al. 1992, McLean et al. 1993, Smith et al. 1996, Durden et al. 1997). The prevalence of *B. burgdorferi* s.s. and the ability of passerines to disperse infected larval and nymphal *Ixodes scapularis* ticks suggests that migratory passerines serve as reservoirs and are involved in the movement and expansion of Lyme disease spirochetes in North America (Lane 1991, Smith et al. 1996, Durden et al. 1997, Rand et al. 1998).

Seabirds spend most of their life in the open ocean, generally nesting in coastal cliffs or offshore islands without mammalian predators to breed over periods of 2–6 mo. After chicks are reared, they return to a pelagic existence until the next breeding season. Seabird colonies in tropical latitudes may be infested by many species of soft ticks (e.g., *Ornithodoros capensis*) as well as hard ticks (e.g., *Ixodes* spp.) (Clifford 1979, Duffy 1991). In northern temperate latitudes in both hemispheres, seabird colonies are infested by *Ixodes uriae*, a widespread and abundant tick species (Zumpt 1952, Eveleigh and Threlfall 1974, Clifford 1979).

Olsen et al. (1993) documented the presence of *B. garinii* from *I. uriae* ticks feeding on Razorbills (*Alca torda*) on the island of Bonden, Sweden. Additionally, the authors found infections from skin biopsies from the Razorbill, providing some of the first evidence of a Lyme disease cycle involving seabirds and *I. uriae* ticks. Since the island was rodent-free, these data also showed that seabirds could serve as competent reservoir hosts without the involvement of mammalian reservoirs. Subsequently, *I. uriae* ticks feeding on Atlantic Puffins (*Fratercula arctica*) on the Faeroe Islands, Black Guillemots (*Cepphus grylle*) in Iceland, and Fork-tailed Storm Petrels (*Oceanodroma furcata*) in Alaska tested positive for *B. garinii* (Olsen et al. 1995, Gylfe et al. 1999). The presence of *B. garinii* in the Southern Hemisphere in King Penguins (*Aptenodytes patagonicus*) and Black-browed Albatrosses (*Diomedea melanophris*) from Campbell Island off New Zealand as well as the Falkland Islands testified to its widespread occurrence in both the hemispheres (Olsen et al. 1995, Gauthier-Clerc et al. 1999).

Many important seabird colonies sustaining globally significant populations occur in the northwest Atlantic (Lock et al. 1994, Gaston and Jones 1998). Recently, the presence of *B. garinii* was recorded from *I. uriae* ticks on Gull Island, Newfoundland, Canada, constituting the first record of this spirochete from a colony in the northwest Atlantic (Smith et al. 2006). Since dispersal and wintering movements of immature and adult seabirds from the North Atlantic (Lock et al. 1994, Gaston and Jones 1998, Huettmann and Diamond 2000) overlap with some of the seabird colonies with recorded *B. garinii* infestations (Olsen et al. 1995, Bunikis et al. 1996), we hypothesized that infections of *B. garinii* were more widespread and present in more seabird species. Additionally, since *B. garinii* is widespread and identical strains occur in both hemispheres, we reviewed the evidence on the phylogeny and ecology of *B. garinii* strains from seabirds to attempt to explain the current distribution of this spirochete among seabirds.

The objectives of this study were (1) to determine the prevalence of *B. garinii* in selected seabird colonies of Newfoundland and Labrador and (2) to review the potential of seabirds and the associated tick species *Ixodes uriae* in the dispersal of *B. garinii* over short and long distances.

MATERIALS AND METHODS

Study Area

Ticks were collected alive from the Gannet Islands, Labrador, in 2005 and 2006; Cape St. Mary's Seabird Sanctuary in 2006; and Gull Island, Newfoundland in 2004, 2005, and 2006. The Gannet Islands Ecological Reserve consists of a group of small islands about 29 km off the coast of Cartwright, southern Labrador (54°00′N, 56°30′W). A cluster of six islands is referred to individually as Gannet Clusters 1 through 6 (GC1–6) (Lock et al. 1994). Five islands (GC1–5) are located within 500 m of one another, with GC6, the largest of the

cluster, located 1.5 km west of the GC1. Highest seabird densities occur during the summer breeding season on GC1–4 and Outer Gannet. The Gannet Islands collectively host over 39,300 breeding pairs of Atlantic Puffins, 10,000 breeding pairs of Razorbills, over 1,270 breeding pairs of Thick-billed Murres (*Uria lomvia*), and over 47,000 breeding pairs of Common Murres (*U. aalge*). The deer mouse (*Peromyscus maniculatus*) is abundant on most of the Gannet Islands. Cape St. Mary's Ecological Reserve is one of six ecological reserves of Newfoundland and Labrador. It is located about 200 km southwest of St. John's on the southwestern tip of the Avalon Peninsula (46°50′N, 54°12′W). About 24,000 Northern Gannets (*Morus bassanus*), 20,000 Black-legged Kittiwakes (*Rissa tridactyla*), 20,000 Common Murres, and 2,000 Thick-billed Murres live within the reserve during the breeding season. Land mammals such as the short-tailed weasel (*Mustela erminea*) and the meadow vole (*Microtus pennsylvanicus*) have access to the colony, which is partly on the mainland.

Gull Island (47°15′N, 52°46′W) is located in southeastern Newfoundland, Canada. It is one of four islands in the Witless Bay Ecological Reserve and is about 5 km southeast of the town of Witless Bay (Robertson et al. 2004). Gull Island hosts diverse seabird breeding colonies including 350,000 breeding pairs of Leach's Storm-Petrels (*Oceanodroma leucorhoa*), 1,600 pairs of Common Murres, 285 breeding pairs of Razorbills, and 4,300 breeding pairs of Black-legged Kittiwakes (Robertson et al. 2004), over 2,600 breeding pairs of Herring Gulls (*Larus argentatus*), and 88 breeding pairs of Great Black-backed Gulls (*L. marinus*; Robertson et al. 2004). Gull Island has the largest North American colony of Atlantic Puffins, estimated at about 140,000 breeding pairs (Robertson et al. 2004). Land mammals are absent from the colony, although there have been periodic reports of one or two minks (*Mustela vison*) that did not survive the winter.

Laboratory Methods

A total of over 1,500 ticks were shipped alive to the Vector-Borne Disease Laboratory, Maine Medical Center, during the years 2005 and 2006. The following analyses, described briefly here, were conducted at the Maine Medical Center. A subset of ticks was dissected and midguts were screened for the presence of spirochetes by fluorescent microscopy using a polyclonal anti-borrelial antibody

(Donahue et al. 1987). DNA was extracted from positive ticks using the Qiagen DNeasy Tissue Kit (Qiagen, Valencia, CA). DNA amplification was performed in a designated room using genus-specific primers that include the partial sequence of rrs-rrla intergenic spacer region, as described by Bunikis et al. (2004) with use of negative controls. Amplification products were visualized on a 1% agarose gel containing 0.5 μg/ml ethidium bromide. At a second laboratory, ticks positive by fluorescent antibody screen were prepared as above for DNA extraction, and PCR was performed using primers directed at the 16s ribosomal DNA. Sequences of amplicons obtained at both laboratories were confirmed to be *B. garinii*

by comparison with known sequences in the Genbank database.

RESULTS

A total of 181 ticks (from all sites) was tested for *B. garinii*, of which a total of 23 ticks (nymphs and females) from Gull Island and the Gannet Islands tested positive (Table 2.1). Specimens from the Gannet Islands in 2005 and Gull Island in 2004 could not be tested since they were dead on arrival at the Maine Medical Center. The prevalence of *B. garinii* differed between both seabird species and years (Table 2.1). Higher prevalence was observed in female ticks than in nymphs in 2005, although

TABLE 2.1

Prevalence of Borrelia garinii *among* Ixodes uriae *ticks tested from different localities in the northwestern North Atlantic.*

Year	Locality	Source	Life stage	Number tested	Number infected	Prevalence (%)
2005	Gull Island	Atlantic Puffin	Nymph	6	1	16.7
			Female	11	4	36.4
		Herring Gull (chick)	Female	8	3	37.5
		Soil in puffin habitat	Nymph	18	1	5.6
			Female	6	1	16.7
2005 totals				49	10	20.4
2006	Gull Island	Atlantic Puffin	Larva	2	0	0.0
			Nymph	3	0	0.0
			Female	29	3	10.3
		Atlantic Puffin (chick)	Female	7	2	28.6
		Black-legged Kittiwake (chick)	Female	6	0	0.0
		Soil in puffin habitat	Larva	1	0	0.0
			Nymph	34	0	0.0
			Female	4	0	0.0
		Common Murre	Female	11	2	18.2
	Gannet Islands	Common Murre	Female	14	2	14.3
		Razorbill (chick)	Female	12	3	25.0
		Soil in puffin habitat	Nymph	1	0	0.0
			Female	8	1	12.5
2006 totals				132	13	9.8
Grand totals				181	23	12.7

the difference was not significant (Fisher's exact test, $P = 0.074$). All nymphs tested in 2006 were negative. In general, ticks collected from Herring Gull chicks and Atlantic Puffin adults and chicks showed the greatest prevalence of *B. garinii*. The lowest prevalence of the spirochete was recorded from ticks collected from soil or litter samples. Overall prevalence (prevalence of infection among all ticks collected from all sources) was significantly higher in 2005 than in 2006 on Gull Island (Fisher's exact test, $P = 0.028$). Comparisons of sections of the genome of the *B. garinii* isolates showed greater similarity to strains collected from the Faeroe Islands, Slovania, and western Siberia than to the North American strains (data not shown).

DISCUSSION

This study provides insight on the prevalence of the recently recorded *B. garinii* in the northwest Atlantic (Smith et al. 2006). The prevalence of *B. garinii* varied between years and between and among seabird species. Atlantic Puffins had consistently high prevalence in both years, although Herring Gull chicks yielded the highest prevalence in 2005.

On both Gull Island and the Gannet Islands, Atlantic Puffins are abundant, occurring in the tens of thousands (Lock et al. 1994, Robertson et al. 2004). Burrowing habits of the species bring them into close association with *I. uriae* ticks. Additionally, on Gull Island, Herring Gulls nest alongside puffins on grassy slopes (Robertson

et al. 2004), bringing this species in close association with ticks, especially in young gulls, which often hide from predators in puffin burrows (Muzaffar and Jones 2007). The higher prevalence of *B. garinii* on Gull Island in 2005 is concomitant with the higher abundance of ticks in that year, suggesting that increased abundance of ticks could lead to increased prevalence of *B. garinii* infections in seabirds (Muzaffar and Jones 2007). Muzaffar and Jones (2007) also noted differences in the feeding preference of nymphs and adult females of *I. uriae*, with nymphs feeding preferentially on puffin chicks and adult females feeding preferentially on adult puffins. Such differences in feeding activity could be crucial in the dynamics of *B. garinii* in seabirds.

Common and Thick-billed Murres are regarded as the preferred hosts of *I. uriae* ticks and finding *B. garinii* in ticks from Common Murres is not surprising (Clifford 1979). Similarly, Razorbills sometimes share nesting habitats with murres, making them suitable tick hosts and thereby candidates for infection by *B. garinii*. Further work is needed to determine the extent to which these species that were sampled opportunistically for ticks are involved in the ecology of *B. garinii*.

The first recorded incidence of *B. garinii* in seabirds on Bonden Island (12 km from the mainland, Olsen et al. 1993) and subsequently on Malgrundet in the Bothnian Gulf of the Baltic Sea likely originated in mainland Europe (Bunikis et al. 1996) (Fig. 2.1). The Bothnian

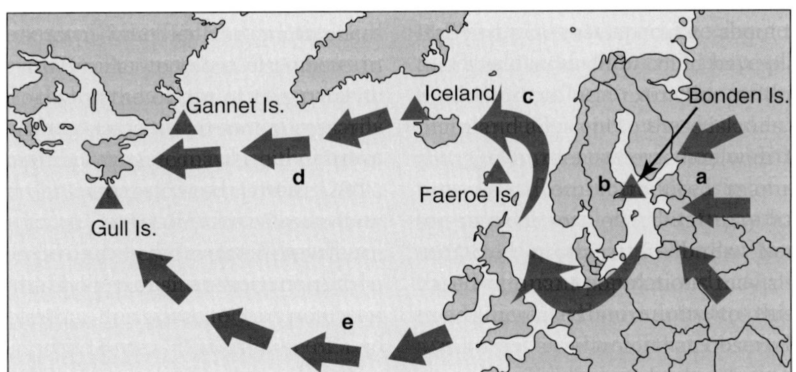

Figure 2.1. Hypothesized movement of *Borrelia garinii* from Europe into seabird colonies in the North Atlantic. (a) Co-occurrence of *Ixodes ricinus* and *I. uriae* on seabird colonies in the Bothnian Gulf and nearby areas. Movement of *B. garinii* from terrestrial to seabird cycle. (b) Establishment of *B. garinii* in seabird colonies along the Northeast Atlantic. (c) Spread of *B. garinii* from endemic focus in Northeast Atlantic to Faeroe Islands and Iceland. (d) Movement of *B. garinii* from Iceland to colonies off the coast of Greenland and Newfoundland. (e) Low-level movement of *B. garinii* infections with birds moving across the Atlantic.

Gulf ecosystem hosts a number of seabird colonies that are close to the mainland and have overlapping populations of *I. uriae* and *I. ricinus* (Clifford 1979, Bunikis et al. 1996). Coastal sites within the Bothnian Gulf off Sweden and Finland, as well as Norway, Denmark, Germany, and the British Isles lining the North Sea, have similar overlapping distributions of these two tick species (Mehl and Traavik 1983, Jaenson et al. 1994). Sympatric distributions present a unique opportunity for *B. garinii* strains from passerines to come in close proximity with *I. uriae*, the vector of *B. garinii* strains in seabirds (Jaenson et al. 1994, Bunikis et al. 1996). Although the two tick species have different ecological niches, their overlapping distributions sometimes permit co-occurrence in similar habitats (Jaenson et al. 1994). Genetically related strains of *B. garinii* have been collected from both these tick species, suggesting a route for the transition from an *I. ricinus–B. garinii* cycle in terrestrial birds and mammals to an *I. uriae–B. garinii* cycle involving seabirds. Co-occurence also suggests that strains of *B. garinii* in the Bothnian Gulf and in the North Atlantic colonies represent a northwestward range expansion of the mainland strains of *B. garinii*, to which they are closely related (Bunikis et al. 1996, Lagal et al. 2003).

Once *B. garinii* had adapted to the seabird transmission cycles involving *I. uriae*, it could then have become established in seabird colonies along the northeast Atlantic colonies through dispersal movements of infected birds between colonies. Establishment could also be facilitated by dispersive movements of *I. uriae* ticks on prospecting fledglings of seabirds, which has been documented in Black-legged Kittiwakes (Danchin 1992, Boulinier and Danchin 1996, McCoy et al. 1999, Boulinier et al. 2001). Subsequently, movements over greater distances could have resulted in the spread of *B. garinii* in the Faeroes and colonies around Iceland (Gylfe et al. 1999). Thick-billed Murres banded in Spitsbergen, for instance, have been recovered from southwest Greenland and Newfoundland (Gaston and Hipfner 2000). Similarly, Razorbills banded as chicks in a colony in Scotland have been found nesting in the Gannet Islands, Labrador (Lavers 2007). With new pockets of endemicity in the eastern North Atlantic, long-dispersal movements of seabirds could then have facilitated the spread of *B. garinii* to colonies in the northwest Atlantic.

Smith et al. (2006) collected ticks from six sites in the northwest Atlantic: Machias Seal Island, Matinicus Rock, Petit Manan Island, and Seal Island in Maine; and Gannet Islands, Labrador, and Gull Island, Newfoundland, in Canada. None of these sites had yielded any evidence of *Borrelia* infections until 2005 on Gull Island (Smith et al. 2006) and 2006 from the Gannet Islands and Gull Island. Previous studies of ticks from a variety of seabirds in the northwest Atlantic colonies, including the Gannet Islands, had failed to find any evidence of *B. garinii* (Gylfe et al. 1999). Tick specimens have been collected earlier and tested for Lyme disease from colonies around Newfoundland (Cape St. Mary's and Gull Island; Whitney 2001) but have never yielded *B. garinii* (Whitney 2001, Bennett 2005). Bennett (2005) sampled 91 *I. uriae* in 2003–2004 from Gull Island, Newfoundland, but these did not yield any evidence of the spirochete.

Borrelia garinii is present in seabird colonies in the North Atlantic. The likely source of infection is from areas of endemicity in the Bothnian Gulf and the northeast Atlantic seabird colonies, where seabirds, songbirds, and two different tick species are in close proximity. Phylogenetic studies suggest a gradual movement of the European strains into seabird colonies in the North Atlantic and then to the northwest Atlantic. Limited information exists on the distribution and movement of *B. garinii* in seabirds. Further studies are urgently needed to help understand patterns of spread and endemism of this spirochete of human health significance. The Gannet Islands colony harbors populations of deer mice that could become adapted to *B. garinii*, and their involvement in the seabird–*Borrelia* cycle needs to be determined. Although *B. garinii* has not yet been documented from Cape St. Mary's, the presence of *I. uriae* ticks and several seabird species in this colony on mainland Newfoundland warrants more detailed examination of this site for this spirochete.

ACKNOWLEDGMENTS

We are indebted to G. Robertson and S. Wilhelm for providing murre specimens from Renews, Newfoundland, and D. Fifield for assistance in collecting tick specimens from Gull Island, Newfoundland. We thank J. Dussureault and B. Hussey for collecting tick specimens from the Gannet Islands, Labrador. We also thank T. Power and his field crew at the

Cape St. Mary's Ecological Reserve for assistance in collecting ticks.

LITERATURE CITED

Anderson, J. F., R. C. Johnson, L. A. Magnarelli, and F. W. Hyde. 1986. Involvement of birds in the epidemiology of the Lyme disease agent *Borrelia burgdorferi*. Infections and Immunity 51:394–396.

Bennett, K. E. 2005. The ticks of insular Newfoundland and their potential for transmitting disease. M.S. thesis, Memorial University of Newfoundland, St. John's, NL, Canada.

Boulinier, T., and E. Danchin. 1996. Population trends in Kittiwakes *Rissa tridactyla* colonies in relation to tick infestation. Ibis 138:326–334.

Boulinier, T., K. D. McCoy, and G. Sorci. 2001. Dispersal and parasitism. Pp. 169–179 *in* J. Clobert, E. Danchin, A. Dhondt, and J. Nichols (editors), Dispersal. Oxford University Press, Oxford, UK.

Bunikis, J., U. Garpmo, J. Tsao, J. Berglund, D. Fish, and A. G. Barbour. 2004. Sequence typing reveals extensive strain diversity of the Lyme borreliosis agents *Borrelia burgdorferi* in North America and *Borrelia afzelii* in Europe. Microbiology 150:1741–55.

Bunikis, J., B. Olsen, V. Fingerle, J. Bonnedahl, B. Wilske, and S. Bergström. 1996. Molecular polymorphism of the Lyme disease agent *Borrelia garinii* in northern Europe is influenced by a novel enzootic *Borrelia* focus in the North Atlantic. Journal of Clinical Microbiology 34:364–368.

Burgdorfer, W., A. G. Barbour, S. F. Hayes, J. L. Benach, E. Grunwaldt, and J. P. Davis. 1982. Lyme disease: A tick-borne spirochetosis? Science 216:1317–1319.

Clifford, C. F. 1979. Tick-borne viruses of seabirds. Pp. 83–100 *in* E. Kurstak (editor), Arctic and tropical arboviruses. Academic Press, London, UK.

Danchin, E. 1992. The incidence of the tick parasite *Ixodes uriae* in Kittiwake *Rissa tridactyla* colonies in relation to the age of the colony and the mechanism of infecting new colonies. Ibis 134:134–141.

Donahue, J. G., J. Piesman, and A. Spielman. 1987. Reservoir competence of white-footed mice for Lyme disease spirochetes. American Journal of Tropical Medicine and Hygiene 36:92–96.

Duffy, D. C. 1991. Ants, ticks and nestling seabirds: dynamic interactions. Pp. 243–257 *in* J. E. Loye, and M. Zuk (editors), Bird-parasite interactions: ecology, evolution and behavior. Oxford University Press, Oxford, UK.

Durden, L. A., R. G. McLean, J. H. Oliver Jr., S. R. Ubico, and A. M. James. 1997. Ticks, Lyme disease spirochetes, trypanosomes and antibody to encephalitis viruses in wild birds from coastal Georgia and South Carolina. Journal of Parasitology 83:1178–1182.

Eveleigh, E. S., and W. Threlfall. 1974. The biology of *Ixodes* (*Ceratixodes*) *uriae* White, 1852 in Newfoundland. Acarologia 16:621–635.

Gaston, A. J., and M. Hipfner. 2000. Thick-billed Murre. A. Poole and F. Gill (editors), The birds of North America No. 497. Academy of Natural Sciences, Philadelphia, PA.

Gaston, A. J., and I. L. Jones. 1998. The Auks. Oxford University Press, Oxford, UK.

Gauthier-Clerc, M., B. Jaulhac, Y. Frenot, C. Bachelard, H. Monteil, Y. Le Maho, and Y. Handrich. 1999. Prevalence of *Borrelia burgdorferi* (the Lyme disease agent) antibodies in King Penguin *Aptenodytes patagonicus* in Crozet Archipelago. Polar Biology 22:141–143.

Gylfe, A., B. Olsen, D. Straševièius, N. M. Ras, P. Weihe, L. Noppa, Y. Ostberg, G. Baranton, and S. Bergström. 1999. Isolation of Lyme disease *Borrelia* from puffins (*Fratercula arctica*) and seabird ticks (*Ixodes uriae*) on the Faeroe Islands. Journal of Clinical Microbiology 37:890–896.

Huettmann, F., and A. W. Diamond. 2000. Seabird migration in the Canadian northwest Atlantic Ocean: moulting locations and movement patterns of immature birds. Canadian Journal of Zoology 78:624–647.

Jaenson, T. G. T., L. Tä Lleklint, L. Lundqvist, B. Olsen, J. Chirico, and H. Mejlon. 1994. Geographical distribution, host associations and vector roles of ticks (Acari: Ixodidae, Argasidae) in Sweden. Journal of Medical Entomology 31:240–256.

Kurtenbach, K., S. De Michelis, S. Etti, S. M. Schäfer, H.-S. Sewell, V. Brade, and P. Kraiczy. 2002. Host association of *Borrelia burgdorferi* sensu lato—the key role of host complement. Trends in Microbiology 10:74–79.

Lagal, V., D. Postic, E. Ruzic-Sabljic, and G. Baranton. 2003. Genetic diversity among *Borrelia* strains determined by single-strand conformation polymorphism analysis of the ospC gene and its association with invasiveness. Journal of Clinical Microbiology 41:5059–5065.

Lane, R. S. 1991. Lyme borreliosis: relation of its causative agent to its vectors and hosts in North America and Europe. Annual Review of Entomology 36:587–609.

Lavers, J. L. 2007. Modeling cumulative mortality and estimating biological parameters for a vulnerable seabird, the Razorbill *Alca torda*, in Atlantic Canada. Ph.D. dissertation, Memorial University of Newfoundland, St. John's, NL, Canada.

Lock, A. R., R. G. B. Brown, and S. H. Gerriets. 1994. Gazetteer of marine birds in Atlantic Canada. Canadian Wildlife Service, Ottawa, ON, Canada.

Magnarelli, L. A., K. C. Stafford III, and V. C. Bladen. 1992. *Borrelia burgdorferi* in *Ixodes dammini* (Acari: Ixodidae) feeding on birds in Lyme, Connecticut, U.S.A. Canadian Journal of Zoology 70:2322–2325.

McCoy, K. D., T. Boulinier, J. Chardine, E. Danchin, and Y. Michalakis. 1999. Dispersal and distribution of the tick *Ixodes uriae* within and among seabird host populations: the need for a population genetic approach. Journal of Parasitology 85:196–202.

McLean, R. G., S. R. Ubico, C. A. N. Hughes, S. M. Engstrom, and R. C. Johnson. 1993. Isolation and characterization of *Borrelia burgdorferi* from blood of a bird captured in the Saint Croix River Valley. Journal of Clinical Microbiology 31:2038–2043.

Mehl, R., and T. Traavik. 1983. The tick *Ixodes uriae* (Acari, Ixodidea) in seabird colonies in Norway. Fauna Norvegica Series B 30:94–107.

Muzaffar, S. B., and I. L. Jones. 2007. Activity periods and questing behavior of the seabird tick *Ixodes uriae* (Acari: Ixodidae) on Gull Island, Newfoundland: the role of puffin chicks. Journal of Parasitology 93:258–264.

Olsen, B., D. C. Duffy, T. G. T. Jaenson, A. Gylfe, J. Bonnedahland, and S. Bergström. 1995. Transhemispheric exchange of Lyme disease spirochetes by seabirds. Journal of Clinical Microbiology 33:3270–3274.

Olsen, B., T. G. T. Jaenson, L. Noppa, J. Bunikis, and S. Bergström. 1993. A Lyme borreliosis cycle in seabirds and *Ixodes uriae* ticks. Nature 362:340–342.

Peisman, J., and L. Gern. 2004. Lyme borreliosis in Europe and North America. Parasitology 129(Supplement):191–220.

Rand, P. W., E. H. Lacombe, R. P. Smith Jr., and J. Ficker. 1998. Participation of birds (Aves) in the emergence of Lyme disease in southern Maine. Journal of Medical Entomology 35:270–276.

Rand, P. W., C. Lubelczyk, G. R. Lavigne, S. Elias, M. S. Holman, E. H. Lacombe, and R. P. Smith Jr. 2003. Deer density and the abundance of *Ixodes scapularis* (Acari: Ixodidae). Journal of Medical Entomology 40:179–184.

Robertson, G. J., S. I. Wilhelm, and P. A. Taylor. 2004. Population size and trends of seabirds breeding on gull and great islands, Witless Bay Ecological Reserve, Newfoundland up to 2003. Canadian Wildlife Service Technical Report Series No. 418. Mount Pearl, NL, Canada.

Smith, R. P., Jr., S. B. Muzaffar, J. Lavers, E. H. Lacombe, B. K. Cahill, C. B. Lubelczyk, and P. W. Rand. 2006. Presence of *Borrelia garinii* in seabird ticks (*Ixodes uriae*) from the Atlantic coast of North America. Emerging Infectious Diseases 12:1909–1912.

Smith, R. P., Jr., P. W. Rand, E. H. Lacombe, S. R. Morris, and D. W. Holmes. 1996. Role of bird migration in the long-distance dispersal of *Ixodes dammini*, the vector of Lyme disease. Journal of Infectious Diseases 174:221–224.

Weisbrod, A. R., and R. C. Johnson. 1989. Lyme disease and migrating birds in the Saint Croix River Valley. Applied Environmental Microbiology 55:1921–1924.

Whitney, H. 2001. Lyme disease in Newfoundland. Animal Health Division, Department of Forest Resources and Agrifoods, Government of Newfoundland and Labrador, Canada.

Zumpt, F. 1952. The ticks of seabirds. Australian National Antarctic Research Expedition Series B:12–20.

CHAPTER THREE

Parasitism in the Endemic Galápagos Dove (*Zenaida galapagoensis*) and Its Relation to Host Genetic Diversity and Immune Response

Diego Santiago-Alarcon, Robert E. Ricklefs,
and Patricia G. Parker

Abstract. Studies on model organisms have shown that individuals with lower genetic diversity are more susceptible to pathogens and suffer greater fitness costs than less inbred individuals. We investigated how genetic diversity, immune response, and parasitism by one hemosporidian parasite (*Haemoproteus multipigmentatus*) and two chewing lice (*Columbicola macrourae* and *Physconelloides galapagensis*) are related to the body condition of endemic Galápagos Doves (*Zenaida galapagoensis*) in six island populations. We hypothesized that (1) host genetic diversity would be negatively related to parasite abundance, (2) genetic diversity would be positively related to body condition of birds, (3) immune response would be positively related to blood parasite intensity but not to louse abundance, (4) alternatively, higher blood parasite intensity generates increased immunosuppression, leading to a lower white blood cell count and indirectly to a lower body condition, and (5) the abundances of the three parasite species would be positively correlated. Genetic diversity measured at eight microsatellite loci was an exogenous variable in the path analysis and explained 58% of the variation in body condition. Our results suggest that genetic diversity directly enhances body condition and indirectly depresses parasite abundance; this pattern was the same for the three parasite species, although it was not significant for *C. macrourae*. The best model suggested that blood parasites increase the activation of the immune system (measured as white blood cell counts), which may indicate an attempt of the host to control infection.

Key Words: avian diseases, Columbiformes, Galápagos, genetic diversity, *Haemoproteus*, immune function, Phthiraptera, *Zenaida galapagoensis*.

Parasitísmo en la Paloma Endémica de Galápagos (*Zenaida galapagoensis*) y su Realción con la Diversidad Genética y la Respuesta Inmune del Huésped

Resumen. Estudios efectuados en organismos de laboratorio han mostrado que los individuos que tienen una menor diversidad genética son más susceptibles a diferentes tipos de patógenos, y al mismo tiempo sufren un mayor costo de salud en comparación a los individuos que presentan mayor diversidad genética. Investigamos como la diversidad genética, la respuesta inmune, y el parasitísmo por un parásito de sangre (*Haemoproteus multipigmentatus*) y dos especies de piójos (*Columbicola macrourae* y *Physconelloides galapagensis*) se relacionan con la condición del cuerpo de las palomas endémicas

Santiago-Alarcon, D., R. E. Ricklefs, and P. G. Parker. 2012. Parasitism in the endemic Galápagos Dove (*Zenaida galapagoensis*) and its relation to host genetic diversity and immune response. Pp. 31–42 *in* E. Paul (editor). Emerging avian disease. Studies in Avian Biology (vol. 42), University of California Press, Berkeley, CA.

de Galápagos (*Zenaida galapagoensis*) en poblaciones de seis diferentes islas. Nuestras hipótesis fueron que (1) la diversidad genética se relacionará negativamente a la abundancia de los tres tipos de parásitos, (2) la diversidad genética se relacionará positivamente a la condición del cuerpo de las aves, (3) la respuesta inmune estaría positivamente correlacionada con la intensidad de los parásitos de sangre, pero no con la abundancia de las dos especies de piójos, (4) alternativamente, una alta intensidad de los parásitos de sangre podría suprimir el sistema inmune lo cual llevaría a un conteo de glóbulos blancos más bajo, e indirectamente a una peor condición del cuerpo, y (5) la abundancia de las tres especies de parásitos estaría positivamente correlacionada. La diversidad genética, medida en base a ocho microsatélites, fue una variable exógena en el análisis de vías y explicó el 58% de la variación en la condición del cuerpo. Los resultados del análisis de vías sugieren que la diversidad genética mejora la condición del cuerpo e indirectamente disminuye la intensidad o abundancia de los parásitos; este patrón fue el mismo para las tres especies de parásitos, aunque no fue significativo para *C. macrourae*. El mejor modelo de análisis de vías sugiere que los parásitos de sangre activan el sistema inmune al incrementar los niveles en los conteos de células blancas, lo cuál puede indicar una reacción del huésped para intentar controlar la infección.

Palabras Clave: Columbiformes, diversidad genética, enfermedades aviar, función inmune, Galápagos, *Haemoproteus*, Phthiraptera, *Zenaida galapagoensis*.

Species inhabiting islands are considered behaviorally and physiologically naïve (Mack et al. 2000). Hawaiian endemic birds represent examples of the problems faced by native species when exposed to disease agents (van Riper et al. 1986, Atkinson et al. 2000). Recently, we have determined that avian endemics in the Galápagos Islands are susceptible to pathogens such as hemosporidian parasites, pox virus, and *Chlamydophila psittaci* (Padilla et al. 2004; Parker et al. 2006; Santiago-Alarcon et al. 2008, 2010). Some of these diseases are not native to the archipelago, such as *C. psittaci* and *Trichomonas gallinae* (Harmon et al. 1987, Parker et al. 2006).

Island species, particularly endemics, tend to harbor lower genetic diversity and have a higher risk of extinction than their continental counterparts (Frankham 1996, 1997, 1998). Genetic diversity has been linked to the evolutionary potential of populations to adapt to challenges imposed by disease agents (Petit et al. 1998, Luong et al. 2007). Thus, factors reducing host genetic diversity can increase susceptibility to diseases. Correlations and other empirical evidence have shown that inbred individuals are more susceptible to parasitism and carry a higher fitness cost when infected compared to non-inbred lines of the same host species (Spielman et al. 2004, Whiteman et al. 2006, Luong et al. 2007). However, few studies have analyzed this relationship in natural populations or related this pattern to immunological responses of the host. Whiteman et al. (2006) studied island populations of the endemic Galápagos Hawk (*Buteo galapagoensis*) and found that the more inbred populations had lower and more variable natural antibody levels. Furthermore, natural antibody levels explained the abundance of the louse *Colpocephalum turbinatum*, which directly interacts with the host immune system when feeding on blood; hawks with higher natural antibody titers had lower abundances of the parasite (Whiteman et al. 2006). Interactions between antibodies and infection were not observed for the louse *Degeeriella regalis*, which feeds only on feathers, dead skin, and keratin and thus does not interact with the host immune system (Whiteman et al. 2006).

Parasites affect the condition of hosts, especially when infection intensities are high (Brown et al. 1995, Merino et al. 2000, Marzal et al. 2005). Feather mass is reduced when lice intensities are high; this can impose an energetic cost that directly impacts host condition via loss of insulative capacity (Booth et al. 1993). The two louse species studied here feed on host feathers and thus interact with mechanical and behavioral defenses rather than the immune system (Moyer et al. 2002). In contrast, blood parasites interact directly with the host immune system (Råberg et al. 2006), and high blood parasite intensity can reduce the survival probability of host individuals, particularly in endemic island birds (Atkinson et al. 2000). Effects on the host by one type of parasite, such as a blood parasite, might facilitate invasion of the host by lice or other types of parasites—an infection with blood parasites can reduce preening activity of the infected bird, creating a potential synergy among parasite effects in multiple infections (Richie 1988).

The Galápagos Islands represent the only Pacific Ocean archipelago that still preserves its entire avifauna (Parker et al. 2006). Some bird populations are declining, however, and we have detected several infectious diseases with interspecific transmission potential in the endemic Galápagos Dove (*Zenaida galapagoensis*; Padilla et al. 2004, Parker et al. 2006). The endemic dove presents high historical gene flow across the archipelago (Santiago-Alarcon et al. 2006); thus, it has a potential role as reservoir for and spreading agent of infectious diseases across dove populations and, depending on the type of disease, to other bird species as well (Parker et al. 2006, Santiago-Alarcon et al. 2006). Here, we studied how genetic diversity, immune response, and parasitism by one hemosporidian parasite (*Haemoproteus multipigmentatus*; Valkiūnas et al. 2010) and two chewing lice (*Columbicola macrourae* and *Physconelloides galapagensis*) are related to the body condition of endemic Galápagos Doves (*Zenaida galapagoensis*). The two louse species analyzed here are native to the archipelago and are specific to the dove (Whiteman et al. 2004); the blood parasites infecting the endemic doves are highly prevalent among island populations and are not endemic to the archipelago; they are rather widespread in continental columbiform populations (Santiago-Alarcon et al. 2010). We measured host body condition, genetic diversity based on microsatellite markers, immune response via white blood cell counts, and parasite abundances in six island populations of the Galápagos Dove. We hypothesized that (1) host genetic diversity would be negatively related to parasite abundance; (2) genetic diversity would be positively related to body condition of birds, potentially indicating superior resistance to parasites of individuals with higher genetic diversity; (3) immune response would be positively correlated to blood parasite intensity but not to louse abundance; (4) alternatively to our third hypothesis, higher blood parasite intensity would generate increased immunosuppression, leading to a lower white blood cell count and indirectly to a lower body condition; and (5) the abundances of the three parasite species would be positively correlated.

METHODS

Field and Lab Work

We conducted this study in the Galápagos archipelago from May through July 2002 on Santiago, Santa Cruz, Santa Fe, and Española Islands; from June through July 2004 on Genovesa Island; and during July 2005 on Wolf Island. We captured endemic doves using hand nets and mist nets following the guidelines in Ralph et al. (1996). For blood parasites, we took blood samples (50 µl) by venipuncture from 25 birds each from Santa Cruz, Santa Fe, and Española Islands; 30 each from Santiago and Genovesa Islands; and 29 from Wolf Island. We visited San Cristóbal Island during 2002 and Darwin Island during 2005, but due to small sample sizes, these islands were not included in our analysis. We prepared two thin blood smears from each Galápagos Dove. Smears were air dried, fixed in methyl alcohol (Dip Quick Fixative; Jorgensen Laboratories Inc., Loveland, CO), and stained with modified Giemsa stain. Blood films were examined for 10–15 min at 40× magnification to detect infection by blood parasites. Intensity of *H. multipigmentatus* was quantified from blood smears by counting the number of parasites observed in 10,000 red blood cells for each infected bird (Valkiūnas et al. 2006). We counted white blood cells (WBC) by examining ten randomly selected fields per slide, also at 40× magnification.

Ectoparasites were quantitatively sampled using the dust-ruffling technique (Walther and Clayton 1997) by applying pyrethroid insecticide (Zema® Flea and Tick Powder for Dogs; St. John Laboratories, Harbor City, CA). Ectoparasites were subsequently stored in vials containing 70% ethanol, and we later counted lice using a stereomicroscope. Dust-ruffling is the method of choice for ectoparasite quantitative sampling when hosts cannot be sacrificed (Clayton and Drown 2001).

Body Condition Index (BCI)

We took the following morphological measurements to the nearest 0.1 mm from the right side of each dove: (1) tarsus from the joint between the tibiotarsus and the tarsometatarsus to the bent joint between the tarsometatarsus and metatarsals; (2) tail from the posterior base of uropygial gland to tip of central rectrices; (3) exposed culmen from the tip of the feathering to the bill's tip; (4) bill width with calipers oriented at a 90° angle to the axis of the bill and the measurement taken at the tip of the feathering; (5) bill depth at the tip of the feathering and at 90° angle to the axis of the bill, and (6) wing chord to the nearest 0.5 mm using a ruler with a perpendicular stop on the unflattened wing from the carpal joint to the tip of the longest primary. Mass was measured to the nearest 0.1 g using 100- and 300-g Pesola scales.

We used principal component analysis (PCA) to obtain a structural size for each individual. We used SPSS (IBM Corp., Armonk NY) in all analyses. Because males are larger than females, analyses were conducted separately for each sex (Santiago-Alarcon and Parker 2007). Although all variables were normally distributed (Kolmogorov–Smirnov test, $P \geq 0.09$) and were measured in the same units except mass, which was not included in the PCA, they were log-transformed to examine the proportional contributions of large and small measurements equally. PCAs were based on a correlation matrix. PC1 for both males and females was the only component with an eigenvalue >1; therefore, we retained PC1 for subsequent estimation of the body condition index (Table 3.1). PC1 explained 64% of the variance in males and 74% of the variance in females; since all measurements loaded positively on PC1 for both sexes, we interpret it as an index of overall body size (Table 3.1). Eigenvectors were rotated using varimax rotation. We retained the rotated eigenvector when the explained variance was higher than that of the unrotated component or when the interpretation of PC1 was more straightforward.

We calculated body condition indices separately for each sex. To calculate the body condition index, we used locally weighted nonlinear regression (LOWESS) to account for the non-linear relationship between body mass and the structural size measurement (PC1) of some of our samples (Green 2001). Residuals of the nonparametric regression (our body condition index) were used in subsequent analyses. PC1 was not correlated with any of the parameters (genetic diversity, leukocyte levels, and parasite abundance) against which our body condition index (i.e., residual mass) was analyzed (see Green 2001 for more details about assumptions and requirements for calculating body condition indices). Analyses were conducted in XLstat add-in for Excel (Addinsoft Inc., New York, NY).

Genetic Analysis

DNA was obtained from blood samples by phenol-chloroform extraction followed by dialysis in 1X TNE2 (10 mM Tris-HCl, 10 mM NaCl, 2 mM EDTA) and diluted to a working concentration of 20 ng/µl. Integrity and concentration of each DNA sample was determined by spectrophotometry and electrophoresis in 0.8% agarose gels run in 1X TBE. Individuals were scored at five polymorphic

TABLE 3.1

Principal component scores (PC1) and communalities (proportion of variance extracted) for each morphological variable for analyses of residual body mass (body condition index) of female and male Galápagos Doves.

Variable	Females		Males	
	PC1	Communalities	PC1	Communalities
Culmen	0.889	0.790	0.825	0.681
Bill width	0.895	0.802	0.693	0.481
Bill depth	0.920	0.846	0.847	0.717
Tarsus	0.863	0.744	0.862	0.743
Tail	0.693	0.480	0.708	0.502
Wing chord	0.884	0.782	0.857	0.735

NOTE: PC scores represent the correlation of each variable with the principal component. Communalities are the squares of the correlation coefficients on the first component or the proportion of variance extracted from each variable.

microsatellite loci (Santiago-Alarcon et al. 2006) and three other loci previously developed for White-winged Doves (*Zenaida asiatica*; S. Tanksley pers. comm.) that were monomorphic for the Galápagos Dove. We prepared 10-μl PCRs that included 50 ng of whole genomic DNA, 1 mM dNTP's, 4μ of 10X reaction buffer, 2μ of 25 mM MgCl2, 0.5 μg of each primer, 0.1 μl of DMSO, and 0.5 units of *Taq* DNA polymerase (SIGMA). PCR conditions were as follows: initial denaturation at 94°C for 3 min, followed by 35 cycles of denaturation at 94°C for 30 sec; annealing from 54–56°C for 1 min, extension at 72°C for 1 min, and a final extension at 72°C for 10 min. PCR products were separated in non-denaturing 7.5% polyacrylamide gels run on BioRad sequencing apparatus. Gels were stained with 0.05% ethidium bromide (EtBr). Gels were visualized using a Kodak UV digital imager (Kodak image station 440CF). Genetic diversity was calculated as a standardized measure of heterozygosity (He) using the program IRmacroN3 for Excel (Amos et al. 2001). We used a PCR-based technique for sexing individuals (Fridolfsson and Ellegren 1999).

Statistical Analyses

Parasites normally exhibit aggregated distributions in host populations, conforming to a negative binomial distribution (Wilson and Grenfell 1997). For this reason, models assuming a normal distribution are not usually appropriate to analyze parasitism data (Wilson et al. 1996, Wilson and Grenfell 1997). It has been shown that general linear models (GLM) assuming a negative binomial error structure are superior to classical linear regression models, even after data have been transformed to fulfill normality requirements (Wilson et al. 1996). Thus, we implemented GLMs in R (version 2.4.1) assuming a negative binomial error for our analyses using the function glm.nb of Venables and Ripley (2002). Model construction was started with a full model and then adjusted by deletion tests (Crawley 2005). Because age of doves (determined by plumage coloration as juvenile or adult; Santiago-Alarcon et al. 2006) and sex were not significant in explaining parasite abundance, we dropped these two variables from subsequent analyses. We constructed models using abundance of parasites as a response variable (one model for each parasite); we controlled for island (fixed factor), and covariates included

heterozygosity, body condition index, and white blood cell counts. We then constructed a similar GLM but used white blood cell counts as the response variable. Results from GLMs produced interaction effects which were difficult to interpret; therefore, we conducted path analyses using structural equation modeling implemented in SPSS Amos (ver. 7.0.0, IBM Corp., Armonk, NY). Path analysis models were constructed following the theoretical base explained in the introduction and guided by results obtained from our GLMs. To select among competing models, we used the Akaike Information Criterion (AIC), and the fit of the model was assessed by using χ^2 test, NFI, CFI, and the RMSEA indices (Klem 2000). Last, we calculated Kendall's rank correlations between parasite abundances, and we corrected the alpha level for multiple comparisons (Bonferroni corrected $\alpha = 0.016$).

RESULTS

General Linear Models

Body condition had a significant negative correlation with parasite abundance for blood parasites and for *P. galapagensis*; the trend was similar for *C. macrourae* but was not significant (Table 3.2). We also identified an interaction effect between body condition and heterozygosity for the two louse species but not for the blood parasites (Table 3.2). We observed an island effect for the three parasite species (Table 3.2). For the GLM with white blood cells as the response variable, blood parasite intensity, but not louse abundance, was a significant effect. We identified a significant interaction between body condition and blood parasite intensity and a three-way interaction between body condition–heterozygosity–blood parasite intensity (Table 3.3); a marginally significant interaction was detected between heterozygosity and blood parasite intensity ($P = 0.10$). Parasite abundances were positively correlated (*H. multipigmentatus–C. macrourae*, $\tau = 0.36$, $P < 0.001$; *H. multipigmentatus–P. galapagensis*, $\tau = 0.29$, $P < 0.001$; *C. macrourae–P. galapagensis*, $\tau = 0.55$, $P < 0.001$).

Path Analysis

To better understand relationships among the different variables, and because the GLMs produced interaction effects that were difficult to interpret,

TABLE 3.2

General linear models for the relationship among parasites, host body condition, and genetic diversity.

Values presented in the table are the z values derived from the model.
Significance levels are indicated as follows: ‡ P = 0.05 to 0.10, * = $P < 0.05$, ** = $P < 0.01$,
*** = $P < 0.001$, ns = $P > 0.10$ and not statistically significant in the model.
Only variables that were significant in any one model are included.

	Response variables		
Explanatory variables	*Columbicola macrourae* abundance	*Physconelloides galapagensis* abundance	*Haemoproteus multipigmentatus* intensity
Intercept	5.63***	4.18***	−6.18***
Island	−3.24**	−5.13***	−4.12***
Body condition index (BCI)	ns	−2.87**	−2.22*
Heterozygosity based on microsatellite markers (He)	1.94‡	1.81‡	ns
BCI × He	1.66‡	3.0**	ns

NOTE: Contrasts were conducted for differences in abundance and intensity among islands. *C. macrourae* abundance: Wolf Island was significantly different from all other islands except Santiago Island; Genovesa Island was significantly different from Santiago Island. *P. galapagensis* abundance: Wolf Island was significantly different from Santa Fe, Santa Cruz, and Genovesa islands; Genovesa Island was significantly different from Santiago Island; Santiago Island was significantly different from Santa Fe and Santa Cruz Islands; Santa Cruz Island was significantly different from Española Island. *H. multipigmentatus* intensity: Genovesa Island was significantly different from all other islands.

TABLE 3.3

General linear model for counts of white blood cells.

Values presented in the table are the z values derived from the model.
Significance levels are indicated as follows: ‡ P = 0.05 to 0.10, * = $P < 0.05$, ** = $P < 0.01$,
*** = $P < 0.001$, ns = $P > 0.10$ and are not statistically significant in the model.
Only variables that were significant in any one model are included.

Explanatory variables	Response variable (white blood cells)
Intercept	2.72**
Body condition index (BCI) × *H. multipigmentatus* intensity	2.12*
Heterozygosity based on microsatellite markers (He) × *H. multipigmentatus* intensity	−1.65‡
BCI × He × *H. multipigmentatus* intensity	−2.11*

we constructed path analysis models based on the theoretical expectations presented in the introduction and using as a guide the results obtained from the GLMs. The model that produced the best fit based on AIC suggested that individual birds with higher genetic diversity had a better body condition (Fig. 3.1). Birds with better body condition also were negatively associated with parasite abundance or intensity; however, the directionality of the effect could not be established (Fig. 3.1). *H. multipigmentatus* intensity had a direct positive effect on white blood cell counts. Reduction of *H. multipigmentatus* intensity by the immune response was not supported by the model. Heterozygosity had significant indirect negative effects on the intensity of *H. multipigmentatus* (−0.21) and on the abundance of *Physconelloides galapagensis* (−0.20) (Fig. 3.1). The black box in the path

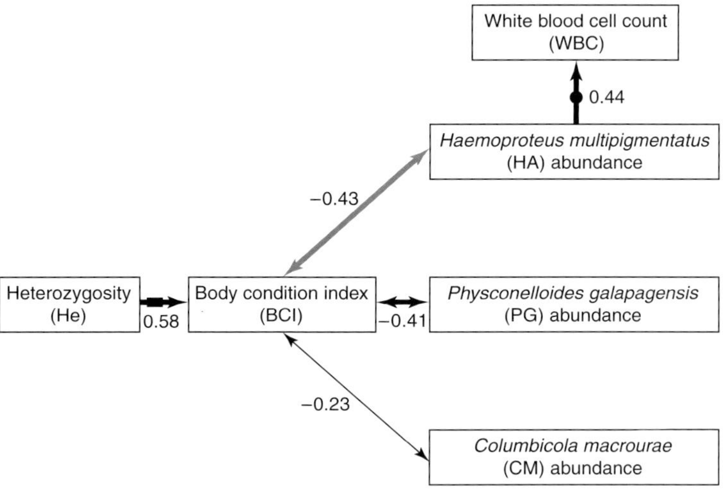

Figure 3.1. Path analysis based on structural equation modeling for the effects of genetic diversity on body condition of birds, parasite abundance, and immune response. $\chi^2 = 64.12$, df $= 10$, $P < 0.001$, NFI $= 0.9$, CFI $= 0.89$, RMSEA $= 0.081$. $\text{AIC}_\text{C} = 382.59$ for best model (combination of hypotheses 1, 2, and 3), $\text{AIC}_\text{C} = 575.49$ for second-best model (combination of hypotheses 2 and 3), $\text{AIC}_\text{C} = 1,005.2$ (combination of hypotheses 1, 2, 3, and 5), and $\text{AIC}_\text{C} = 1,099.5$ for worst model (combination of hypotheses 1, 4, and 5). Black lines represent statistically significant paths ($P < 0.05$), gray lines represent marginally significant paths ($P = 0.05$ to 0.10), and thin black lines represent nonsignificant paths ($P > 0.10$). The black box means several unmeasured intermediate steps (see results). The circle indicates that an alternative hypothesis exists (see results).

diagram indicates that there are several unmeasured steps. For example, neutral genetic variation is unlikely to directly determine the body condition of an individual; it would, rather, have to follow an indirect path to affect the bird's condition (Fig. 3.1). The circle in the path diagram indicates an alternative hypothesis: The positive effect of *H. multipigmentatus* intensity on WBC count may reflect a parallel response of the blood parasites to the presence of unobserved parasites, which may be responsible for eliciting an immune response (Fig. 3.1).

DISCUSSION

In this study, we analyzed the associations of three different parasites (*C. macrourae*, *P. galapagensis*, and *H. multipigmentatus*) with the body condition of the endemic Galápagos Dove and their relationship to host neutral genetic diversity and WBC counts as an index of immune response. Based on a meta-analysis, Reed and Frankham (2003) showed that the fitness of individuals is positively related to their genetic diversity. Links between fitness and diversity are supported by experimental studies of model organisms where more inbred individuals are less resistant to infection and suffer higher fitness costs (Spielman et al. 2004, Luong et al. 2007). Similarly, correlational studies have shown that individuals with higher genetic diversity have lower parasite loads (Acevedo-Whitehouse et al. 2006, Whiteman et al. 2006). Our results indicate that individual doves with higher genetic diversity had better body condition and lower abundances of all three parasite species. Path analysis confirmed the relationship between genetic diversity and body condition but was not able to determine directionality between body condition and parasite abundance. In addition, individuals with higher blood parasite intensities presented higher counts of white blood cells, but two louse species had no significant effect on cell counts. Our results support the hypothesis that parasites that elicit an immunological response are those directly interacting with the immune system of the host (Møller and Rózsa 2005, Whiteman et al. 2006). Thus, our model does not support the hypothesis of immunosuppression, where we would expect a negative effect of blood parasites on counts of white blood cells. Path analysis identified a significant indirect effect of genetic diversity on *P. galapagensis* abundance and on blood parasite intensity. Indirect effects corroborate the triple interaction effect (He × BCI × *H. multipigmentatus* intensity) detected in a GLM where white blood cells were the response variable (Table 3.2). Our results are consistent with one of the main priorities in conservation genetics: the maintenance of species genetic variation.

We did not identify a significant island effect on white blood cell counts. In contrast, Lindström et al. (2004) found a significant island effect of Darwin's finches to ectoparasitic mites and avian pox based on a phytohaemagglutinin injection assay and a keyhole limpet haemocyanin essay. Apanius et al. (2000) found a positive immune response based on WBC to hemosporidian parasites of passerine birds in the Lesser Antilles, which is the pattern that we identified in this study (Ricklefs and Sheldon 2007). Nonetheless, this positive immune reaction due to infection could be produced by other parasites infecting birds and what we observed is just a parallel response of the *H. multipigmentatus* parasites to the immune response elicited by a different parasite(s). Here, we have provided only one measure of immunity (WBC counts); the inclusion of other arms of the immune system likely would reveal trade-offs among different parts of the immune system in response to parasitism (Lindström et al. 2004, Lee et al. 2006). The lack of a direct interaction between genetic diversity and immune response might reflect a trade-off between different parts of the immune system and a rather indirect route by which neutral heterozygosity affects the immune response of an organism, as suggested by the path analysis (Lindström et al. 2004). In addition, we used neutral genetic markers—a global measure of inbreeding—instead of individual markers linked to fitness genes (local effects), where individual markers of local effects may differ from a global measure of genetic diversity in both their direction and occurrence (Lieutenant-Gosselin and Bernatchez 2006). Use of neutral markers might explain the lack of a direct or indirect effect of genetic diversity on the immune response (WBC) in the path analysis model.

Parasite abundances were positively correlated, which suggests a synergistic effect between parasites (Richie 1988). Path analysis, however, did not allow us to discern which parasite is facilitating infection by the other parasites or whether the positive correlations were simply due to parallel responses of the three parasite species to another factor. Furthermore, by adding correlations among parasite species (hypothesis 5), the fit of the model was severely reduced based on AIC_c (see legend of Fig. 3.1). Assuming a synergistic interaction, one possible explanation is that blood parasites make dove individuals more susceptible to infection by the two species of louse. Blood parasites have fitness effects on birds which are particularly severe in endemic birds (Merino et al. 2000; Atkinson et al. 2001a, 2001b; Marzal et al. 2005). Accordingly, the energy that individuals allocate to the immune system to control infections may reduce other behavioral activities such as preening. Preening is the principal way that birds control louse numbers (Clayton 1990, Moyer et al. 2002). Consequently, reduced preening activity due to fitness effects of blood parasites may create an opportunity for lice to increase their population size. Chewing lice and other ectoparasites also reduce the insulation capacity of avian plumage, which increases metabolic rate and reduces body mass and long-term survival of the host (Booth et al. 1993, Brown et al. 1995).

Galápagos Doves are susceptible to *Trichomonas gallinae*, *Chlamydophila psitacci*, and *H. multipigmentatus* (Harmon et al. 1987, Padilla et al. 2004, Santiago-Alarcon et al. 2008). *H. multipigmentatus* prevalence and infection intensities (parasitemia) can be high—100% prevalence in some islands and up to 12% parasitemia (Santiago-Alarcon et al. 2008). High levels of infection intensity are uncommon in natural populations (Valkiūnas 2005). Parasitemias exceeding 12% have been recorded in experimentally infected endemic Hawaiian birds (Atkinson et al. 1995, 2000, 2001a, 2001b; Yorinks and Atkinson 2000). Experimental infections result in high mortality rates, corroborating the high susceptibility of endemic birds to infectious diseases. Results presented in this study suggest that individuals of the endemic dove, even when highly susceptible, may be able to mount an immunological response to challenges imposed by blood parasites if they are in good body condition, which depends on their genetic diversity (Fig. 3.1). The effects identified by our path analysis model should be corroborated with experimental infections that control for the variables analyzed in this study. Our results also suggest that if an exotic pathogen arrives in the archipelago, individuals with less genetic diversity will be more likely to suffer higher fitness consequences. We previously showed that the Galápagos Dove has undergone high historical gene flow across the archipelago (Santiago-Alarcon et al. 2006). High rates of dispersal have conservation implications because the dove can act as reservoir and vector for spread of pathogens which are capable of infecting other species of native birds. In response to potential threats posed by pathogens in the archipelago, our research group, in partnership with the Saint Louis Zoo, the Galápagos National Park, and the Charles Darwin Research Station, established an avian disease surveillance program in 2001 to prevent ecological disasters and to maintain the pristine conditions of the endemic avifauna of the Galápagos (Parker et al. 2006).

ACKNOWLEDGMENTS

For assistance with fieldwork, we thank J. Bollmer, G. Jiménez, J. Merkel, J. Rabenold, A. Iglesias, G. Buitron, and N. Gottdenker. Staff of the Charles Darwin Research Station provided logistical support during the course of this study. N. Cuevas and J. Freire helped with dove sampling at Tortuga Bay, Santa Cruz. Parker's lab group commented on earlier versions of this manuscript. The quality of the paper was greatly improved by comments made by two anonymous reviewers. E. Paul provided suggestions to improve the paper. Permits for sample collection were provided by the Galápagos National Park authorities. TAME airlines provided discount tickets to the Galápagos. Financial support was provided to DS-A and PGP by The Whitney R. Harris World Ecology Center, The Saint Louis Zoo FRC #05-2 and FRC #08-2, and SLZ Wild Care Institute, Frank M. Chapman Memorial Fund of the American Museum of Natural History, Idea Wild, Organization for Tropical Studies, and E. Des Lee Collaborative Vision in Zoological Research. RER was supported by NSF DEB-0542390.

LITERATURE CITED

Acevedo-Whitehouse, K., T. R. Spraker, E. Lyons, S. R. Melin, F. Gulland, R. L. Delong, and W. Amos. 2006. Contrasting effects of heterozygosity on survival and hookworm resistance in California sea lion pups. Molecular Ecology 15:1973–1982.

Amos, W., J. Worthington-Wilmer, K. Fullard, T. M. Burg, J. P. Croxall, D. Bloch, and T. Coulson. 2001. The influence of parental relatedness on reproductive success. Proceedings of the Royal Society of London Series B 268:2021–2027.

Apanius V., N. Yorinks, E. Bermingham, and R. E. Ricklefs. 2000. Island and taxon effects in parasitism and resistance of Lesser Antillean birds. Ecology 81:1959–1969.

Atkinson, C. T., R. J. Dusek, and J. K. Lease. 2001a. Serological responses and immunity to superinfection with avian malaria in experimentally-infected Hawaii Amakihi. Journal of Wildlife Diseases 37:20–27.

Atkinson, C. T., R. J. Dusek, K. L. Woods, and W. M. Iko. 2000. Pathogenicity of avian malaria in experimentally-infected Hawaii Amakihi. Journal of Wildlife Diseases 36: 197–204.

Atkinson, C. T., J. K. Lease, B. M. Drake, N. P. Shema. 2001b. Pathogenicity, serological responses, and diagnosis of experimental and natural malarial infections in native Hawaiian thrushes. Condor 103:209–218.

Atkinson, C. T., K. L. Woods, R. J. Dusek, L. S. Sileo, and W. M. Iko. 1995. Wildlife disease and conservation in Hawaii: pathogenicity of avian malaria (*Plasmodium relictum*) in experimentally infected Iiwi (*Vestiaria coccinea*). Parasitology 111:S59–S69.

Booth, D. T., D. H. Clayton, and B. A. Block. 1993. Experimental demonstration of the energetic cost of parasitism in free-ranging hosts. Proceedings of the Royal Society of London Series B 253:125–129.

Brown, C. R., M. B. Brown, and B. Rannala. 1995. Ectoparasites reduce long-term survival of their avian host. Proceedings of the Royal Society of London Series B 262:313–319.

Clayton, D. H. 1990. Mate choice in experimentally parasitized Rock Doves: lousy males lose. American Zoologist 30:251–262.

Clayton, D. H., and D. M. Drown. 2001. Critical evaluation of five methods for quantifying chewing lice (Insecta: Phthiraptera). Journal of Parasitology 87:1291–1300.

Crawley, M. J. 2005. Statistics: an introduction using R. John Wiley & Sons Ltd., West Sussex, UK.

Frankham, R. 1996. Relationship of genetic variation to population size in wildlife. Conservation Biology 10:1500–1508.

Frankham, R. 1997. Do island populations have less genetic variation than mainland populations? Heredity 78:311–327.

Frankham, R. 1998. Inbreeding and extinction: island populations. Conservation Biology 12:665–675.

Fridolfsson, A. K., and H. Ellegren. 1999. A simple and universal method for molecular sexing of non-ratite birds. Journal of Avian Biology 30:116–121.

Green, A. J. 2001. Mass/length residuals: measures of body condition or generators of spurious results? Ecology 82:1473–1483.

Harmon W. M., W. A. Clark, A. C. Hawbecker, and M. Stafford. 1987. *Trichomonas gallinae* in Columbiform birds from the Galapagos Islands. Journal of Wildlife Diseases 23:492–494.

Klem, L. 2000. Structural equation modeling. Pp. 227–260 *in* L. G. Grimm and P. R. Yarnold (editors), Reading and understanding more multivariate statistics. American Psychological Association, Washington, DC.

Lee, K. A., L. B. Martin II, D. Hasselquist, R. E. Ricklefs, and M. Wikelski. 2006. Contrasting adaptive immune defenses and blood parasite prevalence in closely related *Passer* sparrows. Oecologia 150:383–392.

Lieutenant-Gosselin, M., and L. Bernatchez. 2006. Local heterozygosity-fitness correlations with global positive effects on fitness in threespine stickleback. Evolution 60:1658–1668.

Lindström, K. M., J. Foufopoulos, H. Pärn, and M. Wikelski. 2004. Immunological investments reflect parasite abundance in island populations of Darwin's finches. Proceedings of the Royal Society of London Series B 271:1513–1519.

Luong, L. T., B. D. Heath, and M. Polak. 2007. Host inbreeding increases susceptibility to ectoparasitism. Journal of Evolutionary Biology 20:79–86.

Mack, R. N., D. Simberloff, W. M. Lonsdale, H. Evans, M. Clout, and F. A. Bazzaz. 2000. Biotic invasions: causes, epidemiology, global consequences, and control. Ecological Applications 10:689–710.

Marzal A., F. deLope, C. Navarro, and A. P. Møller. 2005. Malaria parasites decrease reproductive success: an experimental study in a passerine bird. Oecologia 142:541–545.

Merino S., J. Moreno, J. J. Sanz, and E. Arriero. 2000. Are avian blood parasites pathogenic in the wild? A medication experiment in Blue Tits (*Parus caeruleus*). Proceedings of the Royal Society of London Series B 267:2507–2510.

Møller, A. P., and L. Rózsa. 2005. Parasite biodiversity and host defenses: chewing lice and immune response of their avian hosts. Oecologia 142:169–176.

Moyer, B. R., A. T. Peterson, and D. H. Clayton. 2002. Influence of bill shape on ectoparasite load in Western Scrub-Jays. Condor 104:675–678.

Padilla L. R., D. Santiago-Alarcon, J. Merkel, E. Miller, and P. G. Parker. 2004. Survey for *Haemoproteus* spp., *Trichomonas gallinae*, *Chlamidophila pssitaci*, and *Salmonella* spp. in the Galápagos Islands Columbiformes. Journal of Zoo and Wildlife Medicine 35:60–64.

Parker, P. G., N. K. Whiteman, and E. Miller. 2006. Conservation medicine on the Galápagos Islands: partnerships among behavioral, population, and veterinary scientists. Auk 123:625–638.

Petit, R. J., A. El Mousadik, and O. Pons. 1998. Identifying populations for conservation on the basis of genetic markers. Conservation Biology 12:844–855.

Råberg, L., J. C. de Roode, A. S. Bell, P. Stamou, D. Gray, and A. F. Read. 2006. The role of immune-mediated apparent competition in genetically diverse malaria infections. American Naturalist 168:41–53.

Ralph, C. J., G. R. Geupel, P. Pyle, T. E. Martin, D. F. DeSante, and B. Mila. 1996. Manual de métodos de campo para el monitoreo de aves terrestres. USDA Forest Service General Technical Report PSW-GTR-159. USDA Forest Service, Pacific Southwest Research Station, Albany, CA.

Reed, D. H., and R. Frankham. 2003. Correlation between fitness and genetic diversity. Conservation Biology 17:230–237.

Richie, T. L. 1988. Interactions between malaria parasites infecting the same vertebrate host. Parasitology 96:607–639.

Ricklefs, R. E., and K. S. Sheldon. 2007. Malaria prevalence and response to infection in a tropical and in a temperate thrush. Auk 124:1254–1266.

Santiago-Alarcon, D., D. C. Outlaw, R. E. Ricklefs, and P. G. Parker. 2010. Phylogenetic relationships of haemosporidian parasites in New World Columbiformes, with emphasis on the endemic Galapagos dove. International Journal for Parasitology 40:463–470.

Santiago-Alarcon, D., and P. G. Parker. 2007. Sexual size dimorphism and morphological evidence supporting the recognition of two subspecies in the Galápagos Dove. Condor 109:132–141.

Santiago-Alarcon, D., S. M. Tanksley, and P. G. Parker. 2006. Morphological variation and genetic structure of Galápagos Dove (*Zenaida galapagoensis*) populations: issues in conservation for the Galápagos bird fauna. Wilson Journal of Ornithology 118:194–207.

Santiago-Alarcon, D., N. K. Whiteman, P. G. Parker, R. E. Ricklefs, and G. Valkiūnas. 2008. Patterns of parasite abundance and distribution in island populations of endemic Galápagos birds. Journal of Parasitology 94:584–590.

Spielman, D., B. W. Brook, D. A. Briscoe, and R. Frankham. 2004. Does inbreeding and loss of genetic diversity decrease disease resistance? Conservation Genetics 5:439–448.

Valkiūnas, G. 2005. Avian malaria parasites and other haemosporidia. CRC Press. Boca Raton, FL.

Valkiūnas, G., S. Bensch, T. A. Iezhova, A. Križanauskienė, O. Hellgren, and C. V. Bolshakov. 2006.

Nested cytochrome B polymerase chain reaction diagnostics underestimate mixed infections of avian blood haemosporidian parasites: microscopy is still essential. Journal of Parasitology 92:418–422.

Valkiūnas, G., D. Santiago-Alarcon, I. I. Levin, T. A. Iezhova, and P. G. Parker 2010. A new *Haemoproteus* species (Haemosporida: Haemoproteidae) from the endemic Galapagos Dove *Zenaida galapagoensis*, with remarks on the parasite distribution, vectors, and molecular diagnostics. Journal of Parasitology 96:783–792.

van Riper, C., III, S. G. van Riper, M. L. Goff, and M. Laird. 1986. The epizootiology and ecological significance of malaria in Hawaiian land birds. Ecological Monographs 56:327–344.

Venables, W. N., and B. D. Ripley. 2002. Modern applied statistics with S-PLUS. 4th ed. Springer-Verlag, New York, NY.

Walther, B. A., and D. H. Clayton. 1997. Dust-ruffling: a simple method for quantifying ectoparasite loads of live birds. Journal of Field Ornithology 68:509–518.

Whiteman, N. K., K. D. Matson, J. L. Bollmer, and P. G. Parker. 2006. Disease ecology in the Galápagos Hawk (*Buteo galapagoensis*): host genetic diversity, parasite load and natural antibodies. Proceedings of the Royal Society of London Series B 273:797–804.

Whiteman, N. K., D. Santiago-Alarcon, K. P. Johnson, and P. G. Parker. 2004. Differences in straggling rates between two genera of dove lice (Insecta: Phthiraptera) reinforce population genetic and cophylogenetic patterns. International Journal for Parasitology 34:1113–1119.

Wilson, K., and B. T. Grenfell. 1997. Generalized linear modeling for parasitologists. Parasitology Today 13:33–38.

Wilson, K., B. T. Grenfell, and D. J. Shaw. 1996. Analysis of aggregated parasite distributions: a comparison of methods. Functional Ecology 10:592–601.

Yorinks, N., and C. T. Atkinson. 2000. Effects of malaria on activity budgets of experimentally infected juvenile Apapane (*Himatione sanguinea*). Auk 117:731–738.

PART TWO

Population-Level Impacts

Prevalence and Effects of West Nile Virus on Wild American Kestrel (*Falco sparverius*) Populations in Colorado

Robert J. Dusek, William M. Iko, and Erik K. Hofmeister

Abstract. To assess the potential impacts of West Nile virus (WNV) on a wild population of free-ranging raptors, we investigated the prevalence and effects of WNV on American Kestrels (*Falco sparverius*) breeding along the Front Range of the Rocky Mountains in northern Colorado. We monitored kestrel nesting activity at 131 nest boxes from March to August 2004. Of 81 nest attempts, we obtained samples from 111 adults and 250 young. We did not detect WNV in sera; however, 97.3% (108/111) of adults tested positive for WNV neutralizing antibodies. In contrast, 10.0% (23/240) of chicks tested positive for WNV neutralizing antibodies, which possibly represented passive transfer of maternal antibodies. Clutch size, hatching, and fledging success in our study did not differ from that previously reported for this species, suggesting that previous WNV exposure in kestrels did not have an effect on reproductive parameters measured in the breeding population we studied in 2004.

Key Words: American Kestrel, Colorado, *Falco sparverius*, nesting, raptor, reproductive success, West Nile virus.

La Prevalencia y los Efectos del Virus del Nilo Occidental en Poblaciones del Cernícalo Americano (*Falco sparverius*) en Colorado

Resumen. Para determinar los impactos potenciales del virus del Nilo Occidental (VNO) en una población de aves rapaces silvestres, estudiamos la prevalencia y los efectos del VNO en cernícalos americanos (*Falco sparverius*) que nidifican a lo largo del rango frontal de las montañas Rocosas en el norte de Colorado. Monitoreamos la actividad de nidificación en 131 cajas nido de marzo a agosto de 2004. De 81 intentos de nidificación, obtuvimos muestras de 111 adultos y 250 juveniles. No se detectó VNO en el suero; sin embargo, el 97.3% (108/111) de los adultos fueron positivos para los anticuerpos de neutralización del VNO. Por el contrario, el 10.0% (23/240) de los polluelos fueron positivos para los anticuerpos de neutralización del VNO, lo cuál pudo representar la transmisión pasiva de los anticuerpos a través de la madre. El tamaño de la nidada, la eclosión, y el éxito de los volantones durante el estudio no fue diferente a lo que se ha reportado previamente para esta especie, sugiriendo que la exposición previa al VNO en cernícalos americanos no tuvo un efecto sobre los parámetros de reproducción que se midieron en la población reproductiva estudiada en 2004.

Palabras Clave: Cernícalo Americano, Colorado, éxito reproductivo, *Falco sparverius*, nidificación, rapaz, virus del Nilo Occidental.

Dusek, R. J., W. M. Iko, and E. K. Hofmeister. 2012. Prevalence and effects of West Nile virus on wild American Kestrel (*Falco sparverius*) populations in Colorado. Pp. 45–54 *in* E. Paul (editor). Emerging avian disease. Studies in Avian Biology (vol. 42), University of California Press, Berkeley, CA.

The detection of West Nile virus (WNV) in North America raised considerable concern about its effects on wild bird populations. Reports provide evidence that some species have been heavily impacted, including American Crow (*Corvus brachyrhynchos*; Caffrey et al. 2003, Yaremych et al. 2004) and Greater Sage-Grouse (*Centrocercus urophasianus*; Naugle et al. 2004). Other studies analyzing long-term population monitoring data suggest that while some species may be negatively impacted by WNV, others may be unaffected (LaDeau et al. 2007).

Raptors are among the bird groups that are susceptible to WNV. West Nile virus has been detected in at least 34 North American raptor species since its arrival in the Western Hemisphere in 1999 (Nemeth et al. 2006a). While WNV causes morbidity and mortality in numerous species of North American birds, particularly corvids, a health risk to raptors was first documented in 2002. Beginning in the summer and fall of 2002, submissions of sick raptors to rehabilitation centers in the eastern and midwestern U.S. increased substantially (Wünschmann et al. 2004, Joyner et al. 2006, Saito et al. 2007). Investigation into a subset of these submissions concluded that approximately 70% were directly or likely due to WNV infection (Joyner et al. 2006, Saito et al. 2007). In contrast, experimental infections rarely have produced clinical signs of disease or death (Komar et al. 2003; Nemeth et al. 2006a, 2006b). Additionally, WNV antibodies have also been reported from a number of apparently healthy free-living raptors, providing further evidence of their ability to survive infection with this virus (Banet-Noach et al. 2004, Stout et al. 2005, Hull et al. 2006).

In addition to direct mortality brought about by WNV infection, the possibility exists of longer-term effects brought about from infection. Humans with more severe WNV illness can experience fatigue, depression, poor physical health, weakness, and aching that can last for months, and in more severe cases, lifelong neurologic deficits (Rao et al. 2005, Carson et al. 2006, Hayes and Gubler 2006). In birds, this issue is much less understood. Nemeth et al. (2006a) reported a naturally infected Great Horned Owl (*Bubo virginianus*) with mild clinical signs for more than 5 mo while receiving care. Without more detailed studies on free-living and captive raptors, the impacts of this virus relative to raptor populations cannot be completely understood.

To better understand the impacts of WNV on free-living raptors, we initiated a study in 2004 on American Kestrels (*Falco sparverius*) in Colorado. American Kestrels are a common North American raptor with a breeding range that includes much of the continent (Smallwood and Bird 2002). In 2003 in Colorado, kestrels found dead frequently tested positive for WNV (Nemeth et al. 2007) and have previously been reported with WNV antibodies (Medica et al. 2007). However, the short-term and long-term effects of WNV on kestrel populations, including overall survivorship and reproductive success in the wild, are not well understood. In this study, we investigated the prevalence of WNV in a population of American Kestrels; we measured their reproductive success and compared that against baseline data for this species to assess the impacts of (WNV) on a wild population of free-ranging raptors.

METHODS

Nest boxes were monitored from March to August 2004 at multiple sites along the Front Range of the Rocky Mountains in Colorado, from the Denver metropolitan area and north of Fort Collins to the Wyoming border. In the Denver metropolitan area we sampled birds from the following locations: Rocky Mountain Arsenal National Wildlife Refuge (39°49′N, 104°51′W), Barr Lake State Park (39°56′N, 104°45′W), Cherry Creek State Park (39°37′N, 104°50′W), York Street Ponds (39°49′N, 104°57′W), Denver Metro Wastewater Reclamation District (39°48′N, 104°57′W), Riverside Cemetery (39°47′N, 104°57′W), and Aurora Reservoir (39°36′N, 104°39′W). In Northern Colorado we sampled birds at Meadow Springs Ranch (40°54′N, 104°57′W), Rawhide Power Plant (40°51′N, 105°01′W), and in Wellington, Colorado (40°53′N, 105°01′W).

We monitored nest boxes every 10–14 d throughout the breeding season for evidence of nesting activity and to trap and sample adult and nestling kestrels from the box. In addition, we used bal-chatri traps to catch adult birds near nest boxes (McClure 1984, Iko 1991). Nests were checked for activity by closing off the nest box hole and climbing to the box to check for presence of nest cup, eggs, young, or adult kestrels. All adult kestrels were banded with an individually numbered U.S. Geological Survey (USGS) aluminum band and a unique color band combination. Chicks were initially banded with a temporary color band that was removed and replaced with a permanent aluminum USGS band prior to fledging. We also obtained at least one blood sample from all adults and chicks. After sample collection and banding, kestrels were placed back in the box and the nest box hole covered for up to 2 min before reopening. Kestrels captured using a bal-chatri trap were directly released.

At each nest visit the number of eggs or chicks present was recorded. Clutch size was the highest count of eggs made prior to hatching. If hatching did not occur, the nest was recorded as abandoned and that clutch was not incorporated into calculation of mean clutch size. Brood size was determined by direct count of hatched chicks. Fledging success was determined by direct count of chicks that left the nest. If a chick was absent from the nest and known to be ⩾28 d post-hatching and no evidence of chick remains was found in or around the nest, the chick was considered to have fledged. We compared the reproductive parameters from Denver metropolitan area and northern Colorado using Systat 12 (Systat Software Inc., Chicago, IL).

Approximately 1.0 ml of whole blood was collected by jugular or brachial venipuncture from adult birds and transferred to a labeled centrifuge tube with no additives. In some individual adult kestrels, we obtained blood samples on more than one trapping occasion. However, only results from the first blood sample were included in this study. For chicks, blood samples were similarly collected, but of variable volume so that samples did not exceed 1.0% of body weight. We attempted to serially sample kestrel chicks approximately every 10–14 days. Small volume blood samples (⩽0.2 ml) were diluted by putting the sample in a cryovial or centrifuge tube containing 0.5–1.0 ml BA-1 diluent (M199 medium with Hank's salts and Tris HCl [with 7.5% sodium bicarbonate] 20% bovine serum albumin, 20% fetal bovine serum, Penicillin-Streptomycin, 100X, and Fungizone) immediately after collection. All samples were stored with frozen ice packs until processed in the laboratory. For processing, blood samples were separated by centrifugation and sera frozen to −80°C. Serum was later shipped to the USGS National Wildlife Health Center (NWHC) on dry ice, where they remained frozen at −80°C until testing. Serum samples were tested for the presence of specific WNV neutralizing antibodies and determination

of reciprocal antibody titers by plaque reduction neutralization test (PRNT; Beaty et al. 1995). West Nile virus antibody positive samples were also tested for specific St. Louis encephalitis virus (SLE) neutralizing antibodies and reciprocal titer determination by PRNT (Beaty et al. 1995). Serum samples with ⩾90% neutralization of WNV were considered antibody positive. When a serum sample was positive for both WNV and SLE neutralizing antibodies, a four-fold increase in titer of one virus over the other distinguished that virus as the causative etiologic agent of infection resulting in the antibody development. Serum was also tested for WNV by standard plaque assay (Beaty et al. 1995). Appropriate serum, cell, and WNV test dose controls were included in the test.

RESULTS

Of monitored nest boxes, 56% (74/131) were used by kestrels, resulting in 81 nesting attempts—72 first nesting attempts and nine renesting attempts. Two of the first nests were lost from further analysis because one box was knocked down and one box could no longer be checked. Of the remaining first-nest attempts, 74.3% (52/70) successfully fledged at least one young. From these boxes, a total of 111 adults (67 females and 44 males) were captured and sampled for WNV. From the 81 nesting attempts, 260 young hatched, including 103 females, 115 males, and 42 where the sex was unknown because they died or fledged before sex could be determined. Of those that hatched, 224 (86.2%) young successfully fledged.

We obtained serum for all 111 adults and tested for active viral infections and for specific WNV neutralizing antibody. We obtained serum for 241 chicks and tested 240 for WNV specific neutralizing antibody and 224 samples for active viral infections. We were able to obtain an additional 163 samples from recaptured chicks that were also tested for WNV antibody ($n = 162$) and viral infections ($n = 103$). No infectious WNV was detected in 439 (representing 336 individuals) serum samples tested, while all positive control samples yielded virus in the expected quantity. We tested 513 serum samples (representing 351 individuals) for WNV neutralizing antibodies, with 100% (67/67) of the adult females, 93.2% (41/44) of the adult males, and 10.0% (23/240) of chicks testing positive. For adult birds, reciprocal specific WNV neutralizing antibody titers ranged

TABLE 4.1

Specific West Nile virus neutralizing reciprocal antibody titers in American Kestrels
(Falco sparverius), Colorado.

Test subjects	Neg	20	40	80	160	320	640	1,280	2,560
					Titer frequency at first capture				
Adult female ($n = 67$)	0	1	2	4	25	18	13	3	1
Adult male ($n = 44$)	3	1	5	9	11	9	5	1	
Chick male ($n = 113$)	101	8	2		2				
Chick female ($n = 101$)	92	5	3		1				
Chick unknown ($n = 26$)	24	2							

from 20 to 2,560, whereas for chicks the range was 20 to 160 (Table 4.1). The chicks testing positive for WNV antibodies represented nine (15.5%, $n = 58$) different nests where chicks were sampled (Table 4.2).

Nesting success for first nests was not significantly different for any of the parameters measured between Denver metropolitan area and northern Colorado sites (Table 4.3; clutch size, $t = -1.58$, $P = 0.121$; brood size, $t = -0.06$, $P = 0.963$; number of fledglings, $t = -0.75$, $P = 0.458$). For all first nesting attempts, mean clutch size was 4.8 ($n = 58$, SD $= 0.49$), mean brood size 4.3 ($n = 58$, SD $= 0.99$), and mean number of fledglings per successful nest 3.77 ($n = 57$, SD $= 1.58$). Nest abandonment of first nests after at least one egg was laid was 25.7% (18/70).

We recorded renesting attempts in nine nest boxes. In those boxes we captured five banded females that had previously attempted nesting (four in their original box and one in a new box not previously used by any birds in 2004); we did not capture the adult female in two of the boxes; and for the two other boxes it was the second nest recorded for the box but with a new, previously unbanded, adult female. Of the five banded females that renested, four fledged at least one nestling; in three cases this represented a second successful nest. The four that successfully renested had WNV neutralizing antibody titers of 160, 160, 320, and 640; and for the one that did not successfully renest, the titer was 160. The two adult females that represented a second nest in a box previously occupied by another female were unsuccessful nesters and had titers of 320 and 1,280.

DISCUSSION

Our study took place in the spring of 2004, one year following an epizootic and epidemic WNV season in Colorado. We documented a high prevalence of specific WNV neutralizing antibodies in kestrels, but we detected no evidence of ongoing WNV transmission. The high WNV seroprevalence among breeding adult kestrels was likely at least in part a result of the extensive WNV transmission in Colorado in 2003. While WNV was initially detected in Colorado in 2002, in 2003 Colorado had the highest number of WNV human cases in the United States (U.S. Centers for Disease Control, http://www.cdc.gov/ncidod/dvbid/ westnile/surv&control.htm). The first WNV case detected in Colorado in 2004 was on 30 May in a human, suggesting that limited transmission was occurring among wild birds and mosquitoes previous to this date (ProMED-Mail 2004). West Nile virus antibody persistence is poorly studied but has been documented to persist in wild-caught captive Rock Pigeons (*Columba livia*) for >1 yr and in Fish Crows (*Corvus ossifragus*) for at least 1 yr (Gibbs et al. 2005, Yabsley et al. 2007).

American Kestrels found dead have been tested for WNV as part of annual state WNV surveillance programs. In 2000, 57% ($n = 14$) of kestrel carcasses in New York tested positive for WNV (Bernard et al. 2001). In 2003 in Colorado, 43% ($n = 42$) of kestrel carcasses tested positive for

TABLE 4.2

Relationship of American Kestrel (Falco sparverius) *adult- and chick-specific West Nile virus neutralizing antibody titers for family groups with seropositive chicks, Colorado, 2004.*

Nest	Adults			Chicks (first capture)				Chicks (second capture)		
	Sex	Titer	Date	Sex	Date	Weight (grams)	Titer	Date	Weight (grams)	Titer
BL04	F	320	23-Jun	F	17-May	119	20	26-May	131	Neg
	M	180	16-Apr	F	17-May	118	160	26-May	143	Neg
				M	17-May	89	20	26-May	123	Neg
				M	17-May	104	NR	26-May	125	Neg
				M	17-May	99	NR	26-May	121	Neg
CC05	F	320	22-Apr	F	1-Jun	95	20	15-Jun	144	Neg
	M	80	22-Apr	F	1-Jun	83	Neg	15-Jun	132	Neg
				M	1-Jun	77	20	15-Jun	126	Neg
				M	1-Jun	79	160	15-Jun	128	Neg
				M	1-Jun	65	Neg	15-Jun	118	Neg
CC10	F	640	1-Jun	U	12-Jul	97	20			
	M	320	29-Jun	U	12-Jul	79	20			
				U	12-Jul	87	Neg			
				U	12-Jul	89	Neg			
MS70	F	160	27-May	F	8-Jul	59	40	15-Jul	111	Neg
	M	40	15-Jun	M	8-Jul	62	20	15-Jul	99	Neg
				M	8-Jul	47	40	15-Jul	95	Neg

TABLE 4.2 (*continued*)

TABLE 4.2 (CONTINUED)

Nest	Adults			Chicks (first capture)				Chicks (second capture)		
	Sex	Titer	Date	Sex	Date	Weight (grams)	Titer	Date	Weight (grams)	Titer
MS99	F	160	3-May	F	10-Jun	126	20			
	M	80	17-May	F	10-Jun	122	Neg			
				M	10-Jun	109	20			
				M	10-Jun	114	20			
				M	10-Jun	118	Neg			
RMA19NW	F	80	20-Jul	M	4-Aug	82	20	11-Aug	116	Neg
	M	160	6-Jul	M	4-Aug	75	20	11-Aug	119	Neg
RMA29SE	F	320	28-Apr	F	24-May	118	20	7-Jun	134	Neg
				F	24-May	104	40	7-Jun	134	Neg
				M	24-May	96	40	7-Jun	120	Neg
				M	24-May	108	Neg	7-Jun	112	Neg
				M	24-May	74	Neg	7-Jun	114	Neg
RMA35NW	F	320	27-Apr	F	20-May	81	40	7-Jun	120	Neg
	M	80	13-Apr	F	20-May	92	Neg	7-Jun	137	Neg
				F	20-May	95	Neg	7-Jun	134	Neg
				M	20-May	70	160	7-Jun	110	Neg
				M	20-May	91	Neg	7-Jun	120	Neg
RP40	F	160	21-Apr	F	26-May	101	20	9-Jun	131	Neg
	M	640	5-May	F	26-May	79	Neg	9-Jun	123	Neg
				F	26-May	97	Neg	9-Jun	139	Neg
				M	26-May	95	20	9-Jun	127	Neg

NOTE: Titer results expressed as reciprocal titers. Neg = negative. NR = no result (chick not sampled).

Location	n	Clutch size		Brood size		n fledged chicks	
		$\bar{x}$	SD	$\bar{x}$	SD	$\bar{x}$	SD
Denver metropolitan area	37	4.9	0.49	4.3	1.1	3.9	1.52
Northern Colorado	21	4.7	0.46	4.3	0.78	3.6	1.70[a]

[a] $n = 20$.

WNV (Nemeth et al. 2007). However, little information exists on the numbers of free-ranging kestrels that survive WNV infection. Our data indicate that many wild American Kestrels in Colorado survived infection and developed WNV antibodies. Survival of American Kestrels following infection with WNV is also supported by past research. Our results are similar to those reported in a small breeding population of kestrels in Pennsylvania, where 95% (21/22) of American Kestrels were seropositive for WNV (Medica et al. 2007). Kestrels that were experimentally infected with WNV survived with no clinical signs; however, the sample size was small (Nemeth et al. 2006a).

In addition to assessing WNV among breeding adults, we monitored their nestlings for infectious WNV and WNV neutralizing antibodies. In contrast to the high rate of WNV seroprevalence in adults, we detected low seroprevalence rates in chicks. Due to the lack of evidence of recent WNV infection of individuals within the breeding season (e.g., antibodies only detected in relatively young chicks, no seronegative chicks showed subsequent evidence of seroconversion, no chick first identified as positive was still positive at its second capture, and lack of detection of viremia), antibodies in chicks were likely maternally derived. The observation of higher WNV seroprevalence rates in adults versus their offspring has been corroborated by previous studies. Eighty-eight percent of adult Cooper's Hawks (*Accipiter cooperii*) in southeast Wisconsin were seropositive for WNV but only 2.1% of chicks were seropositive (Stout et al. 2005). In addition, 9.2% of nestling Red-tailed Hawks (*Buteo jamaicensis*) and 12% of nestling Great Horned Owls had detectable WNV antibodies within the same study area (Stout et al. 2005). In serial sampling of individual kestrels in our study, we found none of the chicks that were sero-

positive on the first sampling remained positive on subsequent samplings, which is very strong evidence of maternal antibody transfer (Hahn et al. 2006), and also suggests that detection of antibodies in these chicks may be dependent on how early after hatching chicks are sampled. Maternal antibodies were undetectable in most domestic chicken (*Gallus gallus domesticus*) chicks derived from WNV seropositive hens by 28 d post-hatch (Nemeth and Bowen 2007). In addition, hen sera and egg yolks had similar antibody titers at the time of egg laying, but by 1 d post-hatching, chick serum antibody titers had at least a four-fold (and up to 32-fold) reduction below that of their hens, indicating a sharp drop in detectable titers, which continued through 14 d post-hatching (Nemeth and Bowen 2007). Therefore, detection of maternal antibody transfer among free-living raptors may depend on early sampling of chicks. Stout et al. (2005) sampled chicks beginning at 10 d post-hatching, while we attempted to sample chicks at approximately 7 d post-hatching, when maternal antibodies had potentially waned below detectable levels. Even though maternal antibodies may have been undetectable, they could potentially still offer some level of protection if a chick were infected with WNV, as observed by Nemeth and Bowen (2007) in chickens. At 42 d post-hatching in seven experimentally infected chicks that had previously shown maternal antibodies, three failed to become viremic and the remaining four had viremias of later onset and lower peak levels than their seronegative counterparts (Nemeth and Bowen 2007).

Measurements of reproductive success in American Kestrels are variable depending on geographic location of the population; however, throughout the range of American Kestrels, average clutch size is 4.6 eggs, mean brood size is 3.5 young, and mean number of fledglings is 3.3 (Smallwood and Bird 2002). Except that our mean number of fledglings was slightly higher (3.7), our findings were similar. In studies on the Cooper's Hawk, Red-tailed Hawk, and Great Horned Owl, WNV did not appear to have a detectable impact on population numbers (Stout et al. 2005). In contrast, Medica et al. (2007) suggested that WNV could be a contributing factor in an observed loss of breeding kestrels in their study area. Compared with the findings of Smallwood and Bird (2002), our study population of American Kestrels did not appear to be negatively impacted by WNV. Additionally, three successful double broodings were recorded in our study population in 2004, supporting the notion that the reproductive efforts of some breeding adults were not hindered by previous WNV exposure (Toland 1985, Smallwood and Bird 2002).

While we observed many healthy seropositive kestrels in the wild, increases in submissions of symptomatic raptors including kestrels and other species diagnosed with WNV have been documented in rehabilitation facilities (Joyner et al. 2006, Saito et al. 2007). Unfortunately, it is difficult to determine the exact reasons for this dichotomy. These observations may represent individuals with an increased susceptibility to this disease (Stout et al. 2005). An increased susceptibility may be the result of a number of factors, including age, co-infection with other disease agents, genetic variability, and species differences. Other reasons may include variable doses of WNV because not all mosquitoes will inject the same quantity of virus into a bird due to interrupted feeding, or, if kestrels are becoming infected with WNV via ingestion of infected prey (Komar et al. 2003), virus levels in prey may vary. In addition, awareness has been heightened by concern about WNV, and the public may be more likely to notify a rehabilitation center when an apparently diseased bird is observed. State surveillance programs have also actively solicited reports of dead birds, especially corvids and raptors. In the end, it is difficult to determine whether the increases noted are actually due to an increase in the number of sick and dead raptors due to WNV infection or simply an increase in awareness and concern in human observers. Although our study may not fully address this dichotomy, these results provide a better understanding of WNV impacts on free-living, breeding populations of American Kestrels. However, longer-term monitoring of free-ranging bird populations is needed to assess how WNV may impact the overall population in both the short and long term.

ACKNOWLEDGMENTS

We would like to thank our staff and collaborators on this project: H. Mohan, I. LeVan, J. Berven, and A. Skeen; S. Gillihan (Rocky Mountain Bird Observatory), N. Ronan (Rocky Mountain Arsenal National Wildlife Refuge), R. Ryder (Colorado State University), B. Petersen (USDA APHIS), J. Sherpelz and G. Kratz (Rocky Mountain Raptor Program, Colorado State University), W. Andelt (Colorado State University), J. Arington (Cherry Creek State Park), M. Bonnum (Aurora Reservoir), C. Dougal (Riverside Cemetery), D. Meyer (City of Ft. Collins), M. O'Brien (Poudre River Power Authority), R. Rivers (Barr Lake State Park), and F. Smylie (Wellington, CO). We would also like to thank K. Griffin, L. Karwal, and M. Lund for laboratory analysis of samples. C. Franson and N. Nemeth provided helpful comments on earlier versions of this manuscript.

LITERATURE CITED

Banet-Noach, C., A. Y. Gantz, A. Lublin, and M. Malkinson. 2004. A twelve-month study on West Nile virus antibodies in a resident and a migrant species of kestrels in Israel. Vector-Borne and Zoonotic Diseases 4:15–22.

Beaty, B., C. Calisher, and R. Shope. 1995. Arboviruses. Pp. 204–205 *in* E. H. Lennette, D. Lennette, and E. T. Lennette (editors), Diagnostic procedures for viral, rickettsial, and chlamydial infections. 7th ed. American Public Health Association, Washington, DC.

Bernard, K. A., J. G. Maffei, S. A. Jones, E. B. Kauffman, G. D. Ebel, A. P. Dupuis II, K. A. Ngo, D. C. Nicholas, D. M. Young, P. Y. Shi, V. L. Kulasekera, M. Eidson, D. J. White, W. B. Stone, NY State West Nile Virus Surveillance Team, and L. D. Kramer. 2001. West Nile virus infection in birds and mosquitoes, New York State, 2000. Emerging Infectious Diseases 7:679–685.

Caffrey, C., T. J. Weston, and S. C. R. Smith. 2003. High mortality among marked crows subsequent to the arrival of West Nile virus. Wildlife Society Bulletin 31:870–872.

Carson, P. J., P. Konewko, K. S. Wold, P. Mariani, S. Goli, P. Bergloff, and R. D. Crosby. 2006. Long-term clinical and neuropsychological outcomes of West Nile virus infection. Clinical Infectious Diseases 43:723–730.

Gibbs, S. E. J., D. M. Hoffman, L. M. Stark, N. L. Marlenee, B. J. Blitvich, B. J. Beaty, and D. E. Stallknecht. 2005. Persistence of antibodies to West Nile virus in naturally infected Rock Pigeons (*Columba livia*). Clinical and Diagnostic Laboratory Immunology 12:665–667.

Hahn, D. C., N. M. Nemeth, E. Edwards, P. R. Bright, and N. Komar. 2006. Passive West Nile virus antibody transfer from maternal Eastern Screech-Owls (*Megascops asio*) to progeny. Avian Diseases 50:454–455.

Hayes, E. B., and D. J. Gubler. 2006. West Nile virus: epidemiology and clinical features of an emerging epidemic in the United States. Annual Review of Medicine 57:181–194.

Hull, J., A. Hull, W. Reisen, Y. Fang, and H. Ernest. 2006. Variation of West Nile virus antibody prevalence in migrating and wintering hawks in central California. Condor 108:435–439.

Iko, W. M. 1991. Changes in the body mass of American Kestrels (*Falco sparverius*) during the breeding season. M.Sc. degree, University of Saskatchewan, Saskatoon, SK, Canada.

Joyner, P. H., S. Kelly, A. A. Shreve, S. E. Snead, J. M. Sleeman, and D. A. Pettit. 2006. West Nile virus in raptors from Virginia during 2003: clinical, diagnostic, and epidemiologic findings. Journal of Wildlife Diseases 42:335–344.

Komar, N., S. Langevin, S. Hinten, N. Nemeth, E. Edwards, D. Hettler, B. Davis, R. Bowen, and M. Bunning. 2003. Experimental infection of North American birds with the New York 1999 strain of West Nile virus. Emerging Infectious Diseases 9:311–322.

LaDeau, S. L., A. M. Kilpatrick, and P. P. Marra. 2007. West Nile virus emergence and large-scale declines of North American bird populations. Nature 447:710–714.

McClure, E. 1984. Bird banding. Boxwood Press, Pacific Grove, CA.

Medica, D. L., R. Clauser, and K. Bildstein. 2007. Prevalence of West Nile virus antibodies in a breeding population of American Kestrels (*Falco sparverius*) in Pennsylvania. Journal of Wildlife Diseases 43:538–541.

Naugle, D. E., D. L. Aldridge, B. L. Walker, T. E. Cornish, B. J. Moynahan, M. J. Holloran, K. Brown, G. D. Johnson, E. T. Schmidtmann, R. T. Mayer, C. Y. Kato, M. R. Matchett, T. J. Christiansen, W. E. Cook, T. Creekmore, R. D. Falise, E. T. Rinkes, and M. S. Boyce. 2004. West Nile virus: pending crisis for Greater Sage-Grouse. Ecology Letters 7:704–713.

Nemeth, N. M., S. Beckett, E. Edwards, K. Klenk, and N. Komar. 2007. Avian mortality surveillance for West Nile virus in Colorado. American Journal of Tropical Medicine and Hygiene 76:431–437.

Nemeth, N. M., and R. A. Bowen. 2007. Dynamics of passive immunity to West Nile virus in domestic chickens (*Gallus gallus domesticus*). American Journal of Tropical Medicine and Hygiene 76: 310–317.

Nemeth, N., D. Gould, R. Bowen, and N. Komar. 2006a. Natural and experimental West Nile virus infection in five raptor species. Journal of Wildlife Diseases 42:1–13.

Nemeth, N. M., D. C. Hahn, D. H. Gould, and R. A. Bowen. 2006b. Experimental West Nile virus infection in Eastern Screech-Owls (*Megascops asio*). Avian Diseases 50:252–258.

ProMED-Mail. 2004. West Nile virus update 2004—Western Hemisphere (03). Archive number 20040617.1624. <http://www.promedmail.org> (1 June 2009).

Rao, N., D. Char, and S. Gnatz. 2005. Rehabilitation outcomes of 5 patients with severe West Nile virus infection: a case series. Archives of Physical and Medical Rehabilitation 86:449–452.

Saito, E. K., L. Sileo, D. E. Green, C. U. Meteyer, G. S. McLaughlin, K. A. Converse, and D. E. Docherty. 2007. Raptor mortality due to West Nile virus in the United States, 2002. Journal of Wildlife Diseases 43:206–213.

Smallwood, J. A., and D. M. Bird. 2002. American Kestrel (*Falco sparverius*). A. Poole and F. Gill (editors), The birds of North America No. 602. Academy of Natural Sciences, Philadelphia, PA.

Stout, W. E., A. G. Cassini, J. K. Meece, J. M. Papp, R. N. Rosenfield, and K. D. Reed. 2005. Serologic evidence of West Nile virus infections in three wild raptor populations. Avian Diseases 49: 371–375.

Toland, B. R. 1985. Double brooding by American Kestrels in central Missouri. Condor 87:434–436.

Wünschmann, A., J. Shivers, J. Bender, L. Carroll, S. Fuller, M. Saggese, A. van Wettere, and P. Redig. 2004. Pathologic findings in Red-tailed Hawks (*Buteo jamaicensis*) and Cooper's Hawks (*Accipiter cooperii*) naturally infected with West Nile virus. Avian Diseases 48:570–580.

Yabsley, M. J., A. E. Ellis, D. E. Stallknecht, and S. E. J. Gibbs. 2007. West Nile virus antibody prevalence in American Crows (*Corvus brachyrhynchos*) and Fish Crows (*Corvus ossifragus*) in Georgia, U.S.A. Avian Diseases 51:125–128.

Yaremych, S. A., R. E. Warner, P. C. Mankin, J. D. Brawn, A. Raim, and R. Novak. 2004. West Nile virus and high death rate in American Crows. Emerging Infectious Diseases 10:709–711.

CHAPTER FIVE

First Example of a Highly Prevalent but Low-Impact Malaria in an Endemic New Zealand Passerine

PLASMODIUM OF TIRITIRI MATANGI ISLAND BELLBIRDS (*ANTHORNIS MELANURA*)

Rosemary K. Barraclough, Taneal M. Cope, Michael A. Peirce, and Dianne H. Brunton

Abstract. Historical surveys for avian hemosporidian parasites in New Zealand have not revealed substantial prevalence within native birds. However, recent detections of avian malaria (*Plasmodium* spp.) within captive native species have been associated with the death of these birds. Such occurrences have highlighted concerns regarding the possibility of a malaria-associated epizootic event within the New Zealand avifauna similar to that witnessed within Hawaii's naïve native bird populations. In contrast to previous findings, we report the first instance of a high prevalence *Plasmodium* (50%, 39/78) within an endemic New Zealand honeyeater, the Bellbird (*Anthornis melanura*), on Tiritiri Matangi Island. Furthermore, since this prevalence was determined via microscopy, it is likely to be an underestimate of the true parasite prevalence within this population. This Bellbird population is productive and anecdotally among the densest within New Zealand. A small and newly establishing mainland Bellbird population, within 20 km of the Tiritiri population, also exhibited 23% (5/22) prevalence. Size and weight of infected and uninfected birds did not differ significantly. No other hematozoa were detected within sampled Bellbirds. This is the first recorded instance of a common, yet non-lethal, association between an endemic passerine and avian malaria.

Key Words: Anthornis melanura, avian malaria, Bellbird, New Zealand, *Plasmodium*.

El Primer Ejemplo de una Malaria Altamente Prevalente Pero de Bajo Impacto en un Ave Paserina Endémica de Nueva Zelandia: *Plasmodium* en las Campaneras de Nueva Zelandia (*Anthornis melanura*) de la Isla Tiritiri Matangi

Resumen. Los muestreos históricos de los parásitos haemosporida de las aves en Nueva Zelandia no han revelado altas tasas de prevalencia en aves nativas. Sin embargo, recientemente la mortalidad de las especies de aves nativas mantenidas en cautiverio ha sido asociada a infecciones con la malaria aviar (*Plasmodium* spp.). Dichos registros han resaltado las preocupaciones concernientes a la posibilidad de un evento epizoótico de la malaria aviar en la avifauna de Nueva Zelandia, similar al que ocurrió en las poblaciones de aves nativas de Hawaii. A diferencia de hallazgos anteriores, aquí se reporta la primera ocurrencia de una alta tasa de prevalencia de *Plasmodium* (50%, 39/78) en un ave endémica de Nueva Zelandia,

Barraclough, R. K., T. M. Cope, M. A. Peirce, and D. H. Brunton. 2012. First example of a highly prevalent but low-impact malaria in an endemic New Zealand passerine: *Plastmodium* of Tiritiri Matangi Island Bellbirds (*Anthornis melanura*). Pp. 55–64 *in* E. Paul (editor). Emerging avian disease. Studies in Avian Biology (vol. 42), University of California Press, Berkeley, CA.

el Campanero de Nueva Zelandia (*Anthornis mela-nura*), en la Isla Tiritiri Matangi. Ya que la tasa de prevalencia fue determinada usando microscopía, es muy posible que sea una subestimación del verdadero valor de la prevalencia en esta población. La población del Campanero de Nueva Zelandia es productiva y de manera anecdótica esta entre las más densas de Nueva Zelandia. Una pequeña población de esta especie que se estableció recientemente, a 20 km de la población de la Isla Tiritiri, presentó una prevalencia del 23% (5/22). El tamaño y el peso de las aves infectadas y las no infectadas no fue significativamente diferente. Ningún otro parásito Haematozoa fue detectado dentro de las aves muestreadas. Este es el primer registro de una asociación común, pero no letal, entre una paserina endémica y la malaria aviar.

Palabras Clave: Anthornis melanura, Campanero de Nueva Zelandia, malaria aviar, Nueva Zelandia, *Plasmodium.*

The genera *Plasmodium, Haemoproteus,* and *Leucocytozoon* (phylum Apicomplexa, order Haemosporida) are common protozoan parasites of birds. These three genera are closely related (Perkins and Schall 2002, Pérez-Tris et al. 2005) and fall within the same order or family, depending on the author (Garnham 1966, Levine 1988, Valkiūnas 2005).

Prevalence can vary enormously across passerine species worldwide. For example, Deviche et al. (2001) recorded infections ranging from 0 to 90% across 11 breeding Alaskan species. No simple correlation has been found between infection and loss of fitness (Hatchwell et al. 2001). Despite the frequency of these parasites within successful breeding populations, species from each of these three genera have been linked to bird mortalities (Cardona et al. 2002, Schrenzel et al. 2003, Remple 2004).

The most famous instance of pathenogenicity within *Plasmodium* is arguably the Hawaiian *P. relictum capistranoae* epizootic (Warner 1968, van Riper et al. 1986, Atkinson et al. 1995, Freed et al. 2005), where mortality in many endemic species can range from 50 to 90% (Jarvi et al. 2001). The vulnerability of the Hawaiian fauna was assumed to be due to their naïveté, partly attributable to their geographical and evolutionary isolation.

Due to its historic and geographic isolation, New Zealand too has a highly endemic fauna (Daugherty et al. 1993). The forest avifauna of New Zealand, which includes only two international migratory species (two cuckoos, genera *Chrysococcyx* and *Eudynamys*), is also largely ecologically isolated. Consequently, a concern exists regarding the possibility of a Hawaiian-like avian malaria epizootic emerging within New Zealand involving novel parasites (Tompkins and Gleeson 2006). However, New Zealand differs from Hawaii in many important aspects that make it reasonable to expect the presence of native hematozoa along with those established with European settlement (within the last 200 yr). These include its Gondwanan origin and its relative proximity to Australia. For instance, as demonstrated by Adlard et al. (2004), the Australian avifauna carries each of these three genera (11.4% prevalence of one or more hematozoa, $n = 3{,}059$) with generally low levels of parasitemia. Moreover, a high occurrence of *Plasmodium* has been detected elsewhere in the South Pacific, with 59% prevalence found in native land birds of American Samoa (Jarvi et al. 2003). Jarvi et al. (2003) and Atkinson et al. (2006) suggested that this latter Samoan hematozoa fauna is likely to be native due to the low parasitemia, chronic nature of the infections. Furthermore, Atkinson et al. (2006) proposed that the Long-tailed Cuckoo (*Eudynamys taitensis*) that migrates between Samoa and New Zealand might be responsible for shifting parasites into and out of Samoa. In contrast to the above, a thin blood smear survey of 79 Cook Islands birds did not detect any presence of blood parasites (Steadman et al. 1990).

Avian malaria records in New Zealand date back to the early 1900s, when it was predominantly detected in introduced species (Doré 1920a, 1920b, 1921; Laird 1948, 1950). At that time researchers posited malaria as a potential factor in early colonial local native bird extinctions. The possibility of introduced malaria and other avian diseases impacting native birds was also acknowledged in ensuing papers (Turbott 1961). Nevertheless, between Laird's surveys and the work by Fallis et al. (1976), blood parasites are rarely mentioned in the New Zealand literature. The comprehensive thin blood smear work by Fallis et al. (1976) surveyed 43 bird species ($n = 326$ birds), including 113 Passeriformes (among them three Bellbirds) and a single Fiordland Crested Penguin (*Eudyptes pachyrhynchus*) infected with *Leucocytozoon tawaki*. It is seems reasonable to speculate that the absence of intervening reportage was partially due to problems associated with access to protected species, including logistical and legal limitations.

In contrast, recent detections of avian malaria (*Plasmodium* spp.) have been associated with the death of captive birds. In 1996, an outbreak of avian malaria and avian pox in captive endemic New Zealand Dotterel chicks (*Charadrius obscurus*) led to the death of 10 of 16 birds (Richard Jakob-Hoff et al., unpubl. data). More recently an outbreak of malaria among endemic Yellowheads (*Mohoua ochrocephala*) held in a Christchurch wildlife park killed all but one bird in the population (Derraik 2006). Within wild populations, *P. relictum sphenisicdae* had been reported historically from the endangered Yellow-eyed Penguin (*Megadyptes antipodes*; Fantham and Porter 1944, Laird 1950), and it was later suggested that malaria may have caused deaths within this species (Graczyk et al., 1995). Malaria was eventually confirmed in one case of clinical malaria infection by Alley (2001).

Only six further hemosporidian parasite–native host associations have been reported in the New Zealand literature. Recently, a North Island Robin (*Petroica australis*) was found positive for *Haemoproteus* (Parker et al. 2006), whereas *Plasmodium* and *Haemoproteus* have been detected in Saddleback (*Philesturnus carunculatus*) and Tui (*Prosthemadera novaseelandiae*), respectively, in one North Island location, Mokoia Island (Castro and Howe pers. comm.). Laird (1950) identified a single infection of *Plasmodium* sp. in 210 individuals examined of the Grey Duck (*Anas superciliosa*), *P. relictum sphenisicdae* was detected in the Fiordland Crested Penguin (Laird 1950), and an unidentified *Plasmodium* was identified in the New Zealand Pipit (*Anthus novaeseelandiae*) (Doré 1920b, Laird 1950).

The Hawaiian lineage of *Plasmodium* (GRW4, lineage 15; *P. relictum capistranoae*) has been detected within the Australia–Papua New Guinea region (Beadell et al. 2006). *Plasmodium relictum* has also been found in New Zealand (Laird 1950, Tompkins and Gleeson 2006).

Non-hemosporidian parasites identified in New Zealand include an *Atoxoplasma* found within a New Zealand rail, the Stewart Island Weka (*Gallirallus australis scotti*; Laird 1959). The newly described *Babesia kiwiensis* and *Hepatozoon kiwii* have also been described from the North Island Brown Kiwi (*Apteryx australis mantelli*; Peirce et al. 2003).

To advance knowledge of blood parasite prevalence within native New Zealand species, we conducted a thin blood smear survey of the widespread endemic honeyeater, the Bellbird (*Anthornis melanura*), in two locations within the Hauraki Gulf, North Island. The long-standing Tiritiri Matangi Island (Tiritiri) population was the primary target of our survey. The relatively new and self-introduced Bellbird population on the Tawharanui peninsula, nearby (<20 km), was the secondary target.

Bellbirds are relatively abundant in native forests on both main islands of New Zealand, with the exception of the upper North Island, north of Waikato and Northland where they have been locally extinct since the 1860s (Heather and Robertson 1996). Bellbird populations and dispersal have been seriously affected by the removal of native forests and the introduction of invasive predatory species such as cats (*Felis catus*), mustelids (*Mustela*), and rodents (*Rattus*) (Heather and Robertson 1996). However, these predators occur throughout the mainland and the cause of the local extinction of Bellbirds from the north of the North Island is unknown. Bellbirds also inhabit most forested offshore islands, including those in the Hauraki Gulf, alongside the region of mainland extinction. On such predator-free offshore islands such as Tiritiri, Bellbird densities can reach very high numbers; for example, Sagar (1985) measured densities of 22.2–24.4 pairs ha^{-1} on the Poor Knights Islands.

Tiritiri is an island of 220 ha, offshore in the Hauraki Gulf, 28 km north of Auckland. Despite a history of grazing and burning of vegetation (grazed until 1971), a Bellbird population remains on the island. Active restoration of this island started in the 1980s, after which time this population burgeoned. Tawharanui Regional Park is a 588-ha park on an eastern coast peninsula off the North Island that extends into the Hauraki Gulf. The park is a site that receives conservation management for invasive mammalian predators, and its self-introduced Bellbird

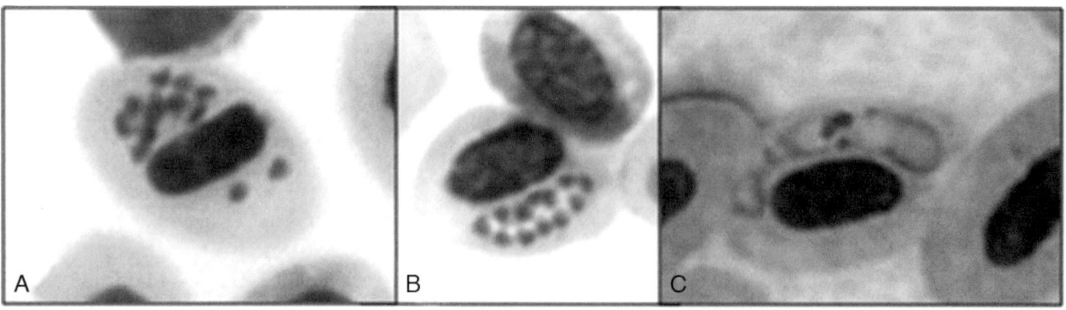

Figure 5.1. Intra-erythrocytic schizonts (a and b) and macrogametocyte (c).

population is growing. The song dialects of this population indicate that Bellbirds are naturally dispersing from Little Barrier Island, also situated in the Hauraki Gulf, rather than Tiritiri (D. H. Brunton, unpubl. data).

METHODS

One hundred Bellbirds were sampled from May 2006 to May 2007, 78 from Tiritiri and 22 from Tawharanui. Tiritiri birds were sampled in conjunction with research on their breeding systems. Birds were mist-netted and captured at feeding stations. Between 5 and 15 µl of blood was taken via venipuncture of the brachial vein. Morphometric measurements (weight, and wing, tail, tarsus, and head–bill lengths) were also taken where possible; however, not every bird was measured.

Thin blood smears were air-dried and fixed with absolute methanol. The smears were stained for 1 hr with Giemsa (AppliChem) at 1:10 dilution. Slides were inspected via light microscopy, initially at 200× for 2 min and then at 1000× oil-immersion for 15 min. Photographs were taken of representative parasites.

Morphometric measurements of Bellbirds positive and negative for *Plasmodium* infection were compared separately for males and females using two sample *t*-tests because male birds are larger than females. Ratios of weight to wing, tarsus, and exposed culmen (correlates of body condition) of Bellbirds positive and negative for *Plasmodium* were also compared separately for males and females using Wilcoxon two-sample tests with pooled data from Tiritiri Island and Tawharanui. No site-dependent differences in morphometrics were known between these two relatively close locations. Statistical analyses were

conducted using SAS (ver 9.11, SAS Institute, Cary, NC) with an alpha level = 0.05.

RESULTS

Plasmodium prevalence within Tiritiri Bellbirds was 50% (39/78), and 22.7% (5/22) in Tawharanui. Overall, Bellbird prevalence was 45%. On available microscopic evidence, the morphology of the parasite is consistent with the subgenus *Novyella*. The number of merozoites observed ranged from 4 to 11 per schizont (Fig. 5.1a–b). The elongated shape of the gametocytes (Fig. 5.1c) confirmed that this *Plasmodium* species is not part of the subgenus *Haemamoeba* complex that includes the invasive Hawaiian *P. relictum capistranoae*. No *Haemoproteus*, *Leucocytozoon*, or other hematozoa were observed during slide inspections.

Tarsus length, head–bill length, wing length, and weight did not vary significantly between infected and non-infected birds. Although the difference between weight and tarsus length of female Bellbirds positive and negative for *Plasmodium* prevalence approached significance ($P = 0.06$, df = 20 and $P = 0.06$, df = 14, respectively; two-sample *t*-test). Wilcoxon tests for ratios between weight and wing length ($P = 0.29$, female, $n = 10$ negative and $n = 5$ positive; $P = 0.16$, male, $n = 27$ negative and $n = 26$ positive), weight:tarsus length ratios ($P = 0.30$, female, $n = 10$ negative and $n = 5$ positive; $P = 0.96$, male, $n = 26$ negative and $n = 27$ positive), and weight:head–bill length ($P = 0.09$, female, $n = 10$ negative and $n = 5$ positive; $P = 0.96$, male, $n = 26$ negative and $n = 27$ positive). Wilcoxon tests also did not detect significant differences between infected birds and those in which parasites were not detected.

Morphometric comparisons were limited by small sample sizes of female Bellbirds. However the *t*-test comparisons between infected male Bellbirds birds and those where no parasites were detected were supported by sample sizes large enough to detect significant differences in morphometric parameters. For example, post hoc power analysis indicates that we had 80% likelihood of detecting a 3-g difference in the weight of female Bellbirds (mean = 23.19 g, negative $n = 14$; mean = 21.53 g, positive $n = 8$ birds), whereas we had a 80% likelihood of detecting a 1.7-g difference in the weight of male Bellbirds (mean = 29.27 g, negative $n = 27$; mean = 29.18, positive $n = 32$ birds).

DISCUSSION

Microscopy does not require an a priori selection of organism-specific primers and is therefore still the standard for survey work when there is no prior knowledge about the diversity of parasites that may be present (Atkinson et al. 2006). However, it is also well established that *Plasmodium* prevalences determined via examination of thin blood smears will tend to underestimate the true prevalence (Jarvi et al. 2003, Ribeiro et al. 2005). Only *Plasmodium* was detected during these slide inspections, and the resulting prevalences likely represent an underestimate of the true *Plasmodium* levels.

It is certain that *Plasmodium* is well established within the prolific Bellbird population on Tiritiri and the new population on Tawharanui. At this stage it is unknown whether the prevalences reported here are similar to those elsewhere in New Zealand, or what the origin of this parasite may be. *Plasmodium* parasites are not necessarily host-specific, and this parasite may be either introduced or native (Szymanski and Lovette 2005, Krizanaskiene et al. 2006). If this *Plasmodium* proves widespread and these levels of prevalence are typical of this host/parasite association, then it is remarkable that this parasite had not been detected previously. However, a number of reasonable potential reasons could explain why it has not been discovered until now. First, the Tiritiri Bellbird population may have a decreased immunocompetence due to historical bottlenecking (as in the North Island Robin; Hale and Briskie 2007) and are therefore likely to carry a higher parasite load. Alternatively, since prevalence levels can change greatly through time, chance may have led to historical thin blood smear surveys coinciding with periods of low prevalence. For example, Bensch et al. (2007) detected a 3–4 yr fluctuation in prevalence of common parasite lineages, sometimes in orders of magnitude.

Another possible explanation for previous lack of detection is that climate and vector-associated differences in prevalence may have led to fewer Bellbird parasite infections at the more southerly, cooler sites of earlier surveys. *Plasmodium* parasites are transmitted to avian hosts by mosquitoes, notably *Culex* and *Aedes* (Raharimanga et al. 2002, Valkiūnas 2005), including *C. quinquefasciatus*, the primary avian malaria vector known from Hawaii, which has been reported to have increased its distribution throughout New Zealand over the last few decades (Tompkins and Gleeson 2006). The degree of exposure to vectors impacts blood-parasite prevalence across habitats and geographical distribution (Rohner et al. 2000, Scheuerlein and Ricklefs 2004, Mendes et al. 2005), and within New Zealand Tompkins and Gleeson (2006) have detected a pattern of decreasing *Plasmodium* prevalence within introduced passerines from north to south, matching the known distribution of *C. quinquefasciatus*. This range in prevalence reflects the north–south winter temperature gradient within New Zealand, and it is reasonable to speculate that any possible nationwide parasitism gradient patterns may be reflected in native as well as introduced bird species.

Whether or not this parasite is native to the Bellbird, one may speculate about the possibility that this *Plasmodium* could have played a part in the historical local extinction of this bird on the mainland adjacent to these islands. For instance, parasite burden interacting with the predation pressures from introduced mammals and the stress from loss of habitat may have negatively impacted northern Bellbirds. Northern island based remnant populations may also have developed increased immunity to this malaria, in a manner similar to that observed within surviving individuals of a vulnerable Hawaiian species, the Amakihi (*Loxops virens*; Atkinson et al. 2001a, Woodworth et al. 2005). Certainly, this *Plasmodium* is not a recent introduction to New Zealand. This is because it has completely penetrated these groups of successfully breeding Bellbirds without any anecdotally observed mortality, as may be associated with the high virulence necessary for quick yet comprehensive invasion.

Breeding success of the Tiritiri Bellbird population for 2005–2006 did not differ significantly from that reported for 1977–1978 (Anderson and Craig 2003, T. Cope, unpubl. data). Given that the Tiritiri Bellbird population continues to be productive and that no significant relationship was found between infection and morphometric measures, it is interesting to speculate on the impact that this parasite may be having on its host. Historically, reports of pathogenicity within the avian Haemosporida were largely confined to a few parasite species that infect domestic birds (Galliformes and Anseriformes; Bennett et al. 1993). However, the comprehensive experimental work done with native species of Hawaiian birds in captivity has illustrated the potential pathogenicity of these parasites for wild birds (Atkinson et al. 1995, Yorinks and Atkinson 2000). Hemosporidian infections have been related to depression in reproduction success (Marzal et al. 2005) and correlated to an inability to mount a strong immune response (Navarro et al. 2003), and infected birds can be more vulnerable to predation (Møller and Nielsen 2007). Furthermore, Atkinson and van Riper (1991) suggest that all species within these genera may be pathogenic, with pathogenicity being related to host specificity, environmental stress, age, nutrition, and the availability of suitable vectors. Last, Peirce et al. (2004) described in detail the physical impact of *Plasmodium*, *Haemoproteus*, and *Leucocytozoon* infections of Australian honeyeaters (Meliphagidae)—for example, schizonts in the spleen, lung, skeletal muscle, liver, and heart displacing functional tissue, as well as local inflammation and fibrous reparation processes. *Plasmodium* spp. were also described causing obstruction of small vessels and inflammation in the spleen, tissue displacement in the lung, and thrombosis in the liver of Noisy Miners (*Manorina melanocephala*). It is also known that the probability of infection increases through time, and adult birds can have higher prevalences due to cumulative exposure (Mendes et al. 2005, Tomé et al. 2005). Despite the fact that no observed die-backs of Bellbirds have been observed on Tiritiri Island, it is possible that small numbers of Bellbirds are being continually lost from the Tiritiri system undetected.

The impact of individual parasite species is unpredictable, as witnessed by the variability in susceptibility of the Hawaiian bird fauna (Atkinson et al. 1995, 2000, 2001b). Even when infection is associated with factors such as poor condition, this may not necessarily negatively affect host survival (Schrader et al. 2003). A variety of studies have found little or no evidence for a negative relationship between intensity of hematozoa parasitism and factors such as host body condition, adult survival, reproductive success, chick growth rates, or recruitment (Dufva 1996, Shutler et al. 1999, Blanco et al. 2001, Bensch et al. 2007). Therefore, no clear relationship exists between infection by these parasites and fitness. Our results did not indicate a significant difference between either male or female infected and uninfected birds. However, it is unfortunate that morphometric parameters were not available for every bird sampled within this study, as this limited the power of our comparisons between infected and uninfected female birds. Since the relationship between both weight and tarsus size and infection status approached significance within females, we suggest that this issue warrants more study within these and other infected Bellbird populations.

What is certain is that these Bellbirds carry a previously unsuspected malarial parasite burden. However, no die-back has been observed within either of these populations over this period of discovery, despite the fact that both these populations are within sites of intensive conservation management by New Zealand Department of Conservation and the Auckland Regional Council. Crucially, this is the first report of a high-prevalence *Plasmodium* population within an endemic New Zealand passerine that is not associated with observed mortality. Thus it is of interest to conservation managers as well as biologists involved in untangling the blood parasite–bird dynamics of New Zealand systems.

ACKNOWLEDGMENTS

Our thanks to the staff of the Auckland Regional Council and the New Zealand Department of Conservation. We also thank B. Gill, J. Ewen, W. Ji, and members of the Ecology and Conservation Group at Massey University, Albany, for their assistance with information and field work.

REFERENCES

Adlard, R. D., M. A. Peirce, and R. Lederer. 2004. Blood parasites of birds from south-east Queensland. Emu 104:191–196.

Alley, M. R. 2001. Avian malaria in Yellow-eyed Penguin. Kokako 8:12–13.

Anderson, S. H., and J. L. Craig. 2003. Breeding biology of Bellbirds (*Anthornis melanura*) on Tiritiri Matangi Island. Notornis 50:75–82.

Atkinson, C. T., R. J. Dusek, K. L. Woods, and W. M. Iko. 2000. Pathogenicity of avian malaria in experimentally infected Hawaii Amakihi. Journal of Wildlife Diseases 36:197–204.

Atkinson, C. T., R. J. Dusek, K. L. Woods, and W. M. Iko. 2001a. Serological responses and immunity to a superinfection with avian malaria in experimentally-infected Hawaii Amakihi. Journal of Wildlife Diseases 37:20–27.

Atkinson, C. T., J. K. Lease, B. M. Drake, and N. P. Shema. 2001b. Pathogenicity, serological responses, and diagnosis of experimental and natural malarial infections in native Hawaiian thrushes. Condor 103:209–218.

Atkinson, C. T., R. C. Utzurrum, J. O. Seamon, A. F. Savage, and D. A. Lapointe. 2006. Hematozoa of forest birds in American Samoa—evidence for a diverse, indigenous parasite fauna from the South Pacific. Pacific Conservation Biology 12:229–238.

Atkinson, C. T., and C. van Riper III. 1991. Pathogenicity and epizootiology of avian haematozoa: *Plasmodium, Leucocytozoon,* and *Haemoproteus.* Pp. 19–48 *in* J. E. Loye and M. Zuk (editors), Bird-parasite interactions. Oxford University Press, New York, NY.

Atkinson, C. T., K. L. Woods, R. J. Dusek, L. S. Sileo, and W. M. Iko. 1995. Wildlife disease and conservation in Hawaii: pathogenicity of avian malaria (*Plasmodium relictum*) in experimentally infected Iiwi (*Vestiaria coccinea*). Parasitology 111:S59–S69.

Beadell, J. S., F. Ishtiaq, R. Covas, M. Melo, B. H. Warren, C. T. Atkinson, S. Bensch, G. R. Graves, Y. V. Jhala, M. A. Peirce, A. R. Rahmani, D. M. Fonseca, and R. C. Fleischer. 2006. Global phylogeographic limits of Hawaii's avian malaria. Proceedings of the Royal Society London Series B 273:2935–2944.

Bennett, G. F., M. A. Peirce, and R. W. Ashford. 1993. Avian hematozoa—mortality and pathogenicity. Journal of Natural History 27:993–1001.

Bensch, S., J. Waldenström, N. Jonzén, H. Westerdahl, B. Hansson, D. Sejberg, and D. Hasselquist. 2007. Temporal dynamics and diversity of avian malaria parasites in a single host species. Journal of Animal Ecology 76:112–122.

Blanco, G., R. Rodríguez-Estrella, S. Merino, and M. Bertellotti. 2001. Effects of spatial and host variables on hematozoa in White-crowned Sparrows wintering in Baja California. Journal of Wildlife Diseases 37:786–790.

Cardona, C. J., A. Ihejirika, and L. McClellan. 2002. *Haemoproteus lophortyx* infection in Bobwhite quail. Avian Diseases 46:249–255.

Daugherty, C. H., G. W. Gibbs, and R. A. Hitchmough. 1993. Mega-island or micro-continent? New Zealand and its fauna. Trends in Ecology and Evolution 8:437–442.

Derraik, J. 2006. Bitten birds: piecing together the avian malaria puzzle. Biosecurity New Zealand 65:16–17.

Deviche, R., E. C. Greiner, and X. Manteca. 2001. Interspecific variability of prevalence in blood parasites of adult passerine birds during the breeding season in Alaska. Journal of Wildlife Diseases 37:28–35.

Doré, A. B. 1920a. Notes on some avian haematozoa observed in New Zealand. New Zealand Journal of Science and Technology 3:10–12.

Doré, A. B. 1920b. The occurrence of malaria in the native Ground Lark. New Zealand Journal of Science and Technology 3:118–119.

Doré, A. B. 1921. Notes on malaria infection in the imported Skylark. New Zealand Journal of Science and Technology 4:126–129.

Dufva, R. 1996. Blood parasites, health, reproductive success, and egg volume in female Great Tits *Parus major.* Journal of Avian Biology 27:83–87.

Fallis, A. M., S. A. Bisset, and F. R. Allison. 1976. *Leucocytozoon tawaki* n.sp. (Eucoccida: Leucocytozoidae) from the penguin *Eudyptes pachyrhynchus,* and preliminary observations on its development in *Austrosimulium* spp. (Diptera: Simuliidae). New Zealand Journal of Zoology 3:11–16.

Fantham, H. B., and A. Porter. 1944. On a Plasmodium (*Plasmodium relictum* var. *spheniscidae* n.var.) observed in four species of penguins. Proceedings of the Zoological Society of London 114:279–292.

Freed, L. A., R. L. Cann, M. L. Goff, W. A. Kuntz, and G. R. Bodner. 2005. Increase in avian malaria at upper elevation in Hawai'i. Condor 107:753–764.

Garnham, P. C. C. 1966. Malaria parasites and other Haemosporidia. Blackwell, Oxford, UK.

Graczyk, T. K., J. F. Cockrem, M. R. Cranfield, J. T. Darby, and P. Moore. 1995. Avian malaria seroprevalence in wild New Zealand penguins. Parasite 2:401–405.

Hale, K. A., and J. V. Briskie. 2007. Decreased immunocompetence in a severely bottlenecked population of an endemic New Zealand bird. Animal Conservation 10:2–10.

Hatchwell, B. J., M. J. Wood, M. A. Anwar, D. E. Chamberlain, and C. M. Perrins. 2001. The haematozoan parasites of Common Blackbirds *Turdus merula*: associations with host condition. Ibis 143:420–426.

Heather, B. D., and H. A. Robertson. 1996. The field guide to the birds of New Zealand. Viking, Auckland, New Zealand.

Jarvi, S. I., C. R. Atkinson, and R. C. Fleischer. 2001. Immunogenetics and resistance to avian malaria in Hawaiian honey creepers (Drepanidinae). Studies in Avian Biology 22:254–263.

Jarvi, S. I., M. E. M. Farias, H. Baker, H. B. Freifeld, P. E. Baker, E. Van Gelder, J. G. Massey, and C. T. Atkinson. 2003. Detection of avian malaria (*Plasmodium* spp.) in native land birds of American Samoa. Conservation Genetics 4:629–637.

Krizanaskiene, A., O. Hellgren, V. Kosarev, L. Sokolov, S. Bensch, and G. Valkiūnas. 2006. Variation in host specificity between species of avian hemosporidian parasites: evidence from parasite morphology and cytochrome B gene sequences. Journal of Parasitology 92:1319–1324.

Laird, M. 1948. Studies on Haematozoa of New Zealand and some adjacent islands. Ph.D. dissertation, University of New Zealand, Wellington. New Zealand.

Laird, M. 1950. Some blood parasites of New Zealand birds. Publications in Zoology from Victoria University College 5:1–20.

Laird, M. 1959. *Atoxoplasma paddae* (Aragão) from several South Pacific silvereyes (Zosteropidae) and a New Zealand rail. Journal of Parasitology 45:47–52.

Levine, N. D., 1988. The protozoan phylum Apicomplexa, Vol. 2, CRC Press, Boca Raton, FL.

Marzal, A., F. de Lope, C. Navarro, and A. P. Møller. 2005. Malarial parasites decrease reproductive success: an experimental study in a passerine bird. Oecologia 142:541–545.

Mendes, L., T. Piersma, M. Lecoq, B. Spaans, and R. E. Ricklefs. 2005. Disease-limited distributions? Contrasts in the prevalence of avian malaria in shorebirds species using marine and freshwater habitats. Oikos 109:296–404.

Møller, A. P., and J. T. Nielsen. 2007. Malaria and risk of predation: a comparative study of birds. Ecology 88:871–881.

Navarro, C., A. Marzal, F. de Lope, and A. P. Møller. 2003. Dynamics of an immune response in House Sparrows *Passer domesticus* in relation to time of day, body condition and blood parasite infection. Oikos 101:291–298.

Parker, K. A., D. H. Brunton, and R. Jakob-Hoff. 2006. Avian translocations and disease: implications for New Zealand conservation. Pacific Conservation Biology 12:155–162.

Peirce, M. A., R. M. Jakob-Hoff, and C. Twentyman. 2003. New species of haematozoa from Apterygidae in New Zealand. Journal of Natural History 37:1797–1804.

Peirce, M. A., R. Lederer, R. D. Adlard, and P. J. O'Donoghue. 2004. Pathology associated with endogenous development of haematozoa in birds from southeast Queensland. Avian Pathology 33:445–450.

Pérez-Tris, J., D. Hasselquist, O. Hellgren, A. Drizanauskiene, J. Waldenström, and S. Bensch. 2005. What are malaria parasites? Trends in Parasitology 21:209–211.

Perkins, S. L., and J. J. Schall. 2002. A molecular phylogeny of malarial parasites recovered from cytochrome b gene sequences. Journal of Parasitology 88:972–978.

Raharimanga, V., F. Soula, M. J. Raherilalao, S. M. Goodman, H. Sadonès, A. Tall, M. Randrianarivelojosia, L. Raharimalala, J. B. Duchemin, F. Ariey, and V. Robert. 2002. Hémoparasites des oiseaux sauvages à Madagascar. Archives de l'Institut Pasteur de Madagascar 68:90–99.

Remple, J. D. 2004. Intracellular hematozoa of raptors: a review and update. Journal of Avian Medicine and Surgery 18:75–88.

Ribeiro, S. F., F. Sebaio, F. C. S. Branquinho, M. A. Marini, A. R. Vago, and E. M. Braga. 2005. Avian malaria in Brazilian passerine birds: parasitism detected by nested PCR using DNA from stained blood smears. Parasitology 130:261–267.

Rohner, C., C. J. Krebs, D. B. Hunter, and D. C. Currie. 2000. Roost site selection of Great Horned Owls in relation to black fly activity: an anti-parasite behavior? Condor 102:950–955.

Sagar, P. M. 1985. Breeding of the Bellbird on the Poor Knights Islands, New Zealand. New Zealand Journal of Zoology 12:643–648.

Scheuerlein, A., and R. E. Ricklefs. 2004. Prevalence of blood parasites in European passeriform birds. Proceedings of the Royal Society London Series B 271:1363–1370.

Schrader, M. S., E. L. Walters, F. C. James, and E. C. Greiner. 2003. Seasonal prevalence of a haematozoan parasite of Red-bellied Woodpeckers (*Melanerpes carolinus*) and its association with host condition and overwinter survival. Auk 120:130–137.

Schrenzel, M. D., G. A. Maalouf, L. L. Keener, and P. M. Gaffney. 2003. Molecular characterization of malarial parasites in captive passerine birds. Journal of Parasitology 89:1025–1033.

Shutler, D., C. D. Ankney, and A. Mullie. 1999. Effects of the blood parasite *Leucocytozoon simondi* on growth rates of anatid ducklings. Canadian Journal of Zoology 77:1573–1578.

Steadman, D. W., E. C. Greiner, and C. Wood. 1990. Absence of blood parasites in indigenous and introduced birds from the Cook Islands, South Pacific. Conservation Biology 4:398–404.

Szymanski, M. M., and I. J. Lovette. 2005. High lineage diversity and host sharing of malarial parasites in a local avian assemblage. Journal of Parasitology 91:768–774.

Tomé, R., N. Santos, P. Cardia, N. Ferrand, and E. Korpimäki. 2005. Factors affecting the prevalence of blood parasites of Little Owls *Athene noctua* in southern Portugal. Ornis Fennica 82:63–72.

Tompkins, D.M., and D. Gleeson. 2006. Relationship between avian malaria distribution and an exotic invasive mosquito in New Zealand. Journal of the Royal Society of New Zealand 36:51–62.

Turbott, E. G. 1961. The interaction of native and introduced birds in New Zealand. Proceedings of the New Zealand Ecological Society 8:62–66.

Valkiūnas G. 2005. Avian malaria parasites and other haemosporida. CRC Press, Boca Raton, FL.

van Riper, C., III, S. G. van Riper, M. L. Goff, and M. Laird. 1986. The epizootiology and ecological significance of malaria in Hawaiian land birds. Ecological Monographs 56:327–344.

Warner, R. E. 1968. The role of introduced diseases in the extinction of the endemic Hawaiian avifauna. Condor 70:101–120.

Woodworth, B. L., C. T. Atkinson, D. A. LaPointe, P. J. Hart, C. S. Speigel, E. J. Tweed, C. Henneman, J. LeBrun, T. Denette, R. DeMots, K. L. Kozar, D. Triglia, D. Lease, A. Gregor, T. Smith, and D. Duffy. 2005. Host population persistence in the face of introduced vector-borne diseases: Hawaii Amakihi and avian malaria. Proceedings of the National Academy of Sciences USA 102:1531–1536.

Yorinks, N., and C.T. Atkinson. 2000. Effects of malaria on activity budgets of experimentally infected juvenile Apapane (*Himatione sanguinea*). The Auk 117:731–738.

Monitoring, Detection, and Research Practices

Prototype System for Tracking and Forecasting Highly Pathogenic H5N1 Avian Influenza Spread in North America

A. Townsend Peterson

Abstract. The recent emergence of a highly pathogenic strain (H5N1) of avian influenza that affects both birds and humans has raised global concern about its spread. Given the rapid spread of the disease and the desire for proactive monitoring and preparedness, I present a prototype forecasting framework for H5N1 dispersal for when/ if it arrives in North America via migratory bird movements. The prototype summarizes movement patterns by six species of arctic-breeding Anseriformes and emphasizes the importance of spread along all coasts of North America, as well as along the lower Mississippi River. This forecasting system is applicable only to the extent that migratory birds are the principal mode of dispersal and spread of the disease. The H5N1 situation calls for considerable effort in (1) understanding details of bird migration globally, (2) sharing avian biodiversity data globally, and (3) exploring novel approaches to data analysis and interpretation.

Key Words: avian influenza, forecasting, H5N1, migratory birds, waterbirds.

Un Sistema Prototipo para Rastrear y Predecir la Expansión de la Influenza Aviar Altamente Infecciosa H5N1 en Norte América

Resumen. La reciente aparición de la cepa altamente patogénica (H5N1) de la influenza aviar que afecta tanto a las aves como a los humanos ha generado una preocupación global acerca de su expansión. Dada la rápida expansión de la enfermedad y el deseo proactivo de un monitoreo y preparación temprana, se presenta el marco de un prototipo para la predicción de la dispersión del H5N1 por medio de los movimientos de las aves migratorias para/por cuando/si llegue/llega en Norte América. El prototipo resume los patrones de movimiento de seis especies de Anseriformes que se reproducen en el Ártico, y enfatiza la importancia de la expansión a lo largo de todas las costas de Norte América, así como también a lo largo de la parte baja del Río Mississippi. El sistema de monitoreo es útil unicamente en situaciones donde las aves migratorias son el principal modo de dispersión y expansión de la enfermedad. La situación del H5N1 requiere de un esfuerzo considerable para (1) entender los detalles de la migración de las aves a nivel global, (2) compartir la información sobre la biodiversidad de las aves a nivel global, y (3) explorar nuevas formas de analizar e interpretar la información.

Palabras Clave: aves acuáticas, aves migratorias, H5N1, influenza aviar, predicción.

Peterson, A. T. 2012. Prototype system for tracking and forecasting highly pathogenic H5N1 avian influenza spread in North America. Pp. 67–80 *in* E. Paul (editor). Emerging avian disease. Studies in Avian Biology (vol. 42), University of California Press, Berkeley, CA.

Flu viruses have long been known to circulate in birds, which led to intensive studies of avian influenzas in the middle of the 20th century. The emergence of a highly pathogenic influenza A strain H5N1—beginning in 1997 in southern China and Southeast Asia, and spreading into Africa, the Middle East, Europe, and even marginally into the Australo-Papuan region (Lavanchy 1998, Yuanji 2002, Tran et al. 2004, Kwon et al. 2005, Normile 2005, Olsen et al. 2006, Parry 2006)—has brought this issue again to the forefront of global attention. The fear is of a global influenza pandemic: This highly pathogenic H5N1 strain (hereafter referred to as "HP-H5N1") is associated with high human case fatality rates, but for the present lacks the ability to be transmitted efficiently among humans (Belshe 2005, Mermel 2005, Hsieh et al. 2006).

For the moment, HP-H5N1 is an avian phenomenon in terms of its transmission cycle; that is, although it is dangerous to humans once they are infected, HP-H5N1 is transmitted efficiently only among birds, and few human-to-human transmission chains have been documented. If, at some point, HP-H5N1 evolves efficient human-to-human transmissibility, then human movements and connectivity will likely govern the spread of the virus (Brockmann et al. 2006). Until then, however, HP-H5N1 is being spread primarily in birds, probably by combinations of dispersal via human-mediated movements of domestic poultry and dispersal by local movements and long-distance migratory movements of wild birds (Kilpatrick et al. 2006). In this paper, I describe initial explorations of using known wild bird movements and patterns of regional connectivity to forecast potential spread patterns within North America.

H5N1 SPREAD

First isolated in 1996 from a farm goose in Guangdong Province, People's Republic of China, HP-H5N1 has spread dramatically since discovery (WHO 2005). The virus was next detected in domestic poultry and humans in Hong Kong in 1997. After a silent period, HP-H5N1 appeared in quick succession in Thailand (December 2003), Republic of Korea (December 2003), Vietnam (January 2004), Japan (January 2004), Thailand (January 2004), Cambodia (January 2004), Laos People's Democratic Republic (January 2004),

Indonesia (February 2004), and People's Republic of China (February 2004). The next set of appearances began with records in People's Republic of China, Indonesia, Thailand, and Vietnam in June and July of 2004, and then in Malaysia (August 2004).

After a pause late in 2004, a large outbreak occurred at Qinghai Lake in central China (April 2005) where 6,000+ wild birds died, as did poultry in Xinjiang Autonomous Region, western China (June 2005), both well to the northwest of previous detections. In quick succession, outbreaks then occurred in western Siberia (Russia, July 2005), Kazakhstan (August 2005), Tibet (August 2005), and Mongolia (August 2005). Finally, late in 2005, HP-H5N1 appeared in Turkey, Romania, Taiwan, and Croatia (all in October 2005), and later in Cyprus, Saudi Arabia, Kuwait, and Iraq (late 2005–early 2006). As of 2007, the virus had been detected across almost all of Europe and the Middle East, in numerous southern and eastern Asian countries (all except Nepal and Bhutan), as well as in several countries in West Africa and the northeastern portion of Africa (Egypt and Sudan; Alexander 2007).

To summarize, from an initial appearance in Southeast Asia, HP-H5N1 spread quickly up the Pacific coast as far as Japan and Korea. It then jumped northward into central Asia, and then appeared to the southwest in the Middle East, Europe, and Africa. These "jumps" have led many to expect increasingly rapid spread, perhaps even globally, in coming years. In reality, though, the apparent extreme rapidity of the spread (particularly in 2005–2006) is most likely a reflection of establishment of surveillance programs—the virus may *already* have spread to most of the areas listed above prior to initiation of surveillance efforts.

ARE WILD BIRDS SPREADING H5N1?

There is little doubt that wild birds are an important long-term reservoir of influenza A viruses (Olsen et al. 2006). However, the broad geographic pattern of spread of HP-H5N1 has been variably interpreted as reflecting involvement of wild birds (FAO 2005, Taubenberger and Morens 2006, Webster et al. 2006) or *not* being consistent with wild bird movements (ABC 2005, National Audubon Society 2005, BirdLife International 2006, JNCC 2006), with a few more balanced opinions (Kilpatrick et al. 2006, Melville and Shortridge 2006). The answer to this debate will guide development of plans for forecasting and remediation in response to HP-H5N1 infections.

The basic argument revolves around the relative weights of four types of evidence: (1) where the virus has been documented to occur, (2) where the virus has *not* been documented to occur, (3) whether the virus has been detected in testing of healthy wild birds, and (4) whether wild birds can be infected with the virus and not become dramatically ill. Brief discussion of these points follows.

Where the Virus Has Been Documented

The pattern of occurrence of HP-H5N1 appears well correlated with wild bird movements. Southeast Asia is the winter home of most migratory birds that breed across East Asia. From there, spread to northeastern Asia (e.g., Japan and Korea) is logical in view of the East Asia/ Australian Flyway. Spread to central Asia (Russia, Mongolia, Kazakhstan) is reasonable given the major waterfowl concentrations at sites such as Qinghai Lake; although these regions are not directly connected by traditionally recognized flyways, movements of some waterbirds (at least in low numbers) to that area from Southeast Asia is highly likely. Last, movement south and west from central Asia into the Middle East, Europe, and Africa is expected given the East Africa/ western Asia flyway.

If we consider a Southeast Asian source area for HP-H5N1, then much of the observed spread of the virus is expected given known migratory bird movements. Colonization of central Asia from Southeast Asia is not in accord with recognized flyways, but movement of some individuals between these areas is likely, at least given the observed levels of straying and vagrancy reported in other regions (BirdLife International 2006, Kilpatrick et al. 2006, Melville and Shortridge 2006). Apart from this discrepancy, the zigzag, north-and-south movements of HP-H5N1 coincide with expectations of ties to migratory bird movements (Peterson et al. 2004).

The alternative explanation centers generally on the poultry industry and its human-mediated movements of birds among regions (BirdLife International 2006). This explanation (at least in isolation) would explain the observed spread patterns much less well. For example, southeastern China is much more "connected" by domestic bird trade with eastern China (e.g., Shanghai, Beijing) than with Japan and Korea, for instance, yet HP-H5N1 has not yet been detected in eastern China. Similarly, trade and movements of poultry between Southeast Asia and Tibet, Mongolia, Russia, Niger, and Kazakhstan cannot be significant. The exclusively human- and poultry-mediated explanation simply does not appear to be a viable explanation (Kilpatrick et al. 2006).

Where the Virus Has Not Been Documented

Another point on which argument has centered has been that of where HP-H5N1 has *not* occurred. The argument is that if the virus is present in central Asia, and if migratory birds are spreading it around, then it should have been detected in India and in Africa, because those regions receive large numbers of migratory birds from Central Asia (BirdLife International 2006). Curiously, since the publication of the BirdLife position paper, HP-H5N1 has been documented in India, Pakistan, Bangladesh, and broadly across West and northeastern Africa. Other regions that have been argued to be curiously *absent* for HP-H5N1 include Taiwan (where it subsequently appeared), the Philippines, and northern and western Australia (FAO 2005).

Two points, however, should be borne in mind. First, in some of these regions of apparent absence, human densities (and, more importantly, poultry densities) are low. As such, nondetection in parts of northern Australia or East Africa should be viewed with caution. Second, HP-H5N1 transmission by migratory birds is clearly a relatively rare event, as its spread has been so spotty and sparse; as such, one would not *expect* it to appear in all areas—the sample of transmission events is small, and some gaps are sure to be present.

Nondetection of the Virus in Wild Birds

Nondetection of the virus in samples from healthy wild birds in Hong Kong, Mongolia, New Zealand, Australia, and Canada has been the focus of considerable comment (BirdLife International 2006). Here, the epidemiology of the disease needs to be taken into account: HP-H5N1 is known to be highly pathogenic to birds, and circulating infections or individuals that have survived infections should rarely be encountered. Nonetheless, it only takes a single actively infected individual to start an outbreak, so low probabilities make detection difficult. Similar difficulties of detection were observed in West Nile virus (WNV) in North America—early initial testing of thousands of migratory birds failed to detect active infections; nonetheless, detailed subsequent studies documented extensive exposure of migratory birds to the virus, allowing them to spread WNV widely in the Western Hemisphere (Dupuis et al. 2003, 2005; Komar 2003; Komar et al. 2003, 2005; Marra et al. 2003).

The BirdLife argument also ignores several studies that *have* detected HP-H5N1 infections in wild birds. For instance, a recent study based on sampling 13,000+ migratory birds in China detected HP-H5N1 eight times (Chen et al. 2006). A recent detailed and extensive study found avian influenza infections (other strains, albeit at low prevalences) in five wild ducks (*Anas* spp.; Winker et al. 2007). Many additional records of HP-H5N1 infections in wild birds can be found in the ProMed archives (www.promedmail.org).

Can Wild Birds Be Infected Asymptomatically?

A last point made as an argument against a role for migratory birds spreading HP-H5N1 (BirdLife International 2006)—that infected birds would be too sick to migrate, and thus could not serve as efficient vectors of HP-H5N1 spread— requires additional study and careful thought. A recent study concluded that at least domestic ducks can indeed be infected asymptomatically, based on the observation that experimentally infected ducks can shed virus for extended periods of time (Hulse-Post et al. 2005), a result confirmed recently in a larger study (Chen et al. 2006). Chen et al. (2006) reported HP-H5N1 isolation from apparently healthy wild ducks from China; more importantly, that study reported 34 of 1,092 (3.1%) migratory ducks seropositive for HP-H5N1, indicating that numerous migratory birds had been infected and had survived the infection.

Thus, the available evidence indicates that migratory waterfowl can indeed be infected with the virus and not necessarily all die or become too sick to migrate. Further insights might be gained via mass testing of healthy birds from areas in which the virus is known to be circulating. Results cited from Canada (BirdLife International 2006) may have little relevance to the prevalence of HP-H5N1 since the virus has not yet spread to the Americas! It is worth noting that early in the WNV invasion of North America, this same point was made repeatedly, suggesting that wild birds simply could not be agents of broad spread of WNV because it killed all the birds it infected— this argument has clearly been counterindicated (Dupuis et al. 2003, Komar et al. 2003, Peterson et al. 2004).

Moreover, long experience with some of the worst infectious diseases and associated mortality rates demonstrates that the idea of universal mortality is untenable, and seeding a new outbreak only takes a single infected individual. For example, Ebola virus ranks among the most dramatic and drastic of diseases that affect humans, and mortality rates range from 40% to 80% (Murphy et al. 1990). Rabies virus is usually cited as universally fatal, but a number of cases of human survival are known (Hemachudha et al. 2002). The point here is that genetic variation is universal, and disease resistance is generally present in at least some individuals of a population—the idea of universal mortality is untenable, and seeding a new outbreak only takes a single infected individual.

Avian Role in Spreading HP-H5N1

A key conclusion regarding HP-H5N1 spread is identification of a need for further information. However, considerable evidence argues in favor of migratory birds playing an important role, although poultry-mediated spread events may also occur. The recent study based on massive samples from China identified phylogenetic connections between viruses isolated from ducks from northern China and Hong Kong, indicating migratory bird transport over ~1,700 km (Chen et al. 2006). As such, the argument that wild birds are "victims not vectors" (BirdLife International 2006) is based on negative evidence only, and is unlikely to be tenable once additional information is available. Rather, wild birds are likely to prove to be both victims *and* vectors in the spread of HP-H5N1.

An additional consideration is the taxonomic distribution of HP-H5N1 infections. A widespread view is that this virus mainly infects waterfowl, but this point may eventually prove to be more dogma than scientific result (Olsen et al. 2006, Taubenberger and Morens 2006, Webster et al. 2006). Waterfowl tend to congregate in large numbers at freshwater and coastal sites, and opportunities for transmission from individual to individual are great. When large-bodied waterbirds get sick or die, carcasses are much more obvious than remains of small-bodied songbirds in forests or other complex habitats.

In the course of studies at several of the HP-H5N1 outbreak sites, landbirds have been found to be infected with the virus. For example, a 29 January 2006 ProMed post documented an Oriental Magpie-Robin (*Copsychus saularis*) found dead in Hong Kong that tested positive for H5N1. A small sample of Eurasian Tree Sparrows (*Passer montanus*) in China showed surprisingly high prevalences of HP-H5N1 (Kou et al. 2005). Several studies now in progress are sampling avifaunas more broadly in an effort to characterize the true host distribution of avian influenza viruses, and results are beginning to appear. As such, the conclusion that waterfowl alone are the hosts of HP-H5N1 (cf. a recent review in Olsen et al. 2006) should probably be reexamined in the face of new evidence from broader sampling programs. If only waterbirds are sampled for HP-H5N1 infection (most avian influenza sampling programs focus exclusively or principally on Anseriformes and Charadriiformes), then one may well conclude that only waterfowl carry these viruses! These considerations should be weighed in designing forecasting systems based on migratory patterns of bird species.

H5N1 IN NORTH AMERICA

I have argued that wild birds play a significant role in spreading HP-H5N1 geographically. While alternative hypotheses are possible, a significant role of wild bird migration in HP-H5N1 spread seems undeniable, and explorations of the implications of bird migration for this phenomenon are merited. Not only can predictions be made regarding HP-H5N1 spread in coming months and years, but also the bird migration–based

scenarios can provide a baseline expectation against which observed patterns can be compared versus causation by other agents such as transport of domestic poultry.

A point of particular interest is linkages among continents. As mentioned in a previous review of WNV dynamics (Peterson et al. 2004), distances in the high Arctic are not great (particularly for migratory birds), and North America and Eurasia are more interconnected by migratory birds than one might expect. A suite of Asian birds extends their breeding range from eastern Siberia to western Alaska, and numerous American birds use both eastern Siberia and westernmost Alaska (Cramp and Perrins 1977–1994). Similar linkages exist between Europe and eastern Canada, mostly centered around Greenland. Last, many pelagic seabirds show long-distance movements that could also provide connectivity among continents. Intercontinental connections were reviewed and analyzed by Peterson et al. (2007), and demonstrate ample connectivity by migratory birds between Eurasia and the Americas. Different sectors of the migratory avifauna may transport the virus to the extreme northwest, the extreme northeast, or broadly along coastal areas of North and even South America.

FORECASTING H5N1 SPREAD IN NORTH AMERICA: A PROTOTYPE

Considering that the Americas may now be one of few regions in which time still remains for strategic planning of responses to and monitoring for these viruses, an understanding of likely "next steps" of the virus, once it arrives in North America, becomes key. For example, if the virus were to appear in Labrador in August, where would it be expected to be in December, if migratory birds were vectors of dispersal?

The principal motivation for this work was the idea that breeding and wintering distributions of migratory bird species are likely highly structured and interconnected (Webster et al. 2002), as has long been known in the monographic literature (Palmer 1976). Birds from the eastern end of a breeding distribution may not have the same probability of ending up at the eastern or western end of the wintering distribution compared to western birds. Techniques for characterizing this non-random connectivity based on fine-resolution genetic information (Kimura et al. 2002) and stable isotope distributions (Hobson 1999, Lott et al. 2003, Farmer et al. 2004, Hobson et al. 2004, Atkinson et al. 2005) are beginning to be explored (Webster et al. 2002) but are not yet sufficiently mature to yield detailed answers in as short a time span as the avian influenza situation may demand; still, these techniques clearly will provide important new insights in the future. Radio-tracking or satellite-based tracking offers impressive detail on individual movements but does not as yet produce sufficient sample sizes to allow detailed and comprehensive conclusions (Berthold et al. 1995).

Here, I describe a prototype of how *existing* diverse information sets (Fig. 6.1) can be used to develop a useful forecasting system now. Data incorporated in this prototype include the North American Breeding Bird Survey and Canadian Breeding Bird Census to establish breeding-season distributional areas, and Christmas Bird Count and other winter-season data sets to establish overwintering areas. Data associated with specimens in natural history museums offer parallel information for Mexico and other regions poorly covered by national monitoring programs. Banding recoveries from the North American Bird Banding Laboratory data set were used to establish connectivity of distributional areas by migratory movements. Last, detailed geospatial data sets were used to permit inference and interpolation of distributional areas. My prototype model treats only southward migratory movements from along the arctic rim of North America, and is based on only six waterfowl species—but it serves to illustrate the concept as well as to teach some interesting lessons about the structure of North American bird migration. The model is already in advanced phases of expansion and improvement (Peterson et al. 2009).

Methods

The prototype model presented here focuses on five species of geese and one seaduck—Ross's

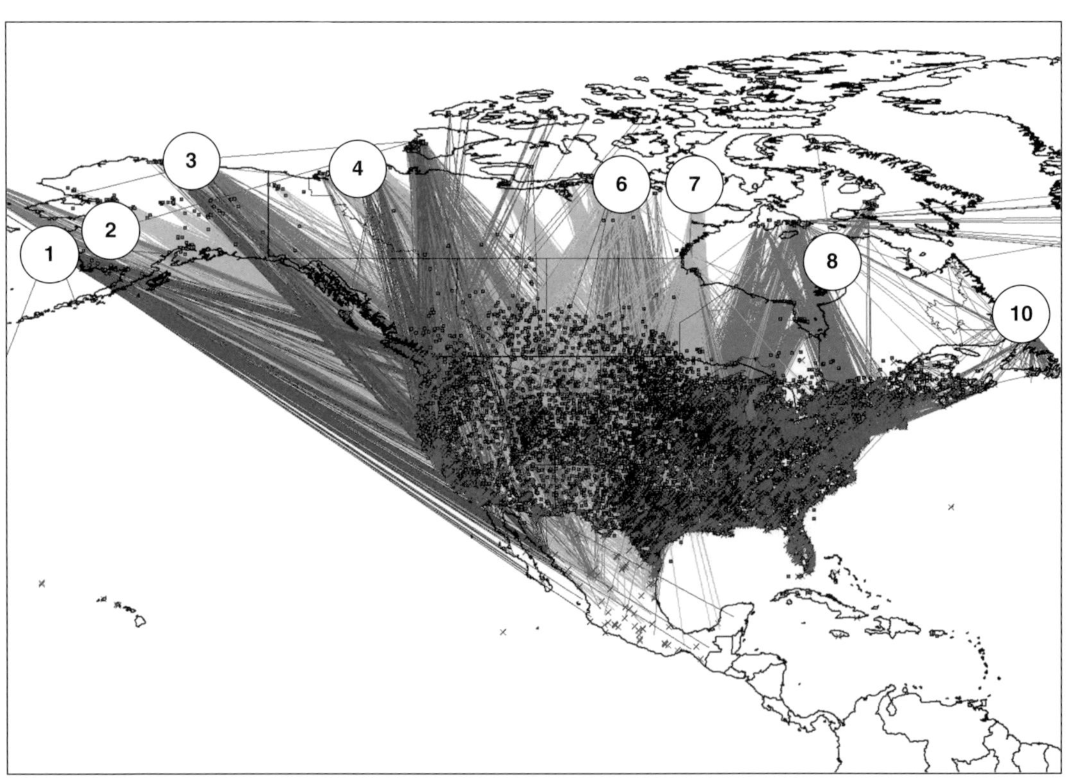

Figure 6.1. Illustration of the large quantity and diversity of avian occurrence data used in the prototype. Shown are points indicating occurrence data from North American Breeding Bird Survey and Christmas Bird Count; lines represent banding-recovery events between breeding and wintering seasons by six species of Anseriformes, with gray shade depicting a different species. The circles along the northern rim of the continent represent the eight exemplar breeding-distribution source areas treated in the prototype. Note the eastbound and westbound banding data linking North America with Greenland and Siberia.

Goose (*Chen rossii*), Snow Goose (*C. caerulescens*), Greater White-fronted Goose (*Anser albifrons*), Atlantic Brant (*Branta bernicla*), Black Brant (*B. nigricans*), and Common Eider (*Somateria mollissima*). The set of species was chosen on the basis of availability of banding data as well as to illustrate high arctic bird migration dynamics, and is not intended to be comprehensive or even necessarily representative. Rather, it is intended to be illustrative of the challenge and of the complexities of bird migration across North America.

Raw occurrence data were taken from the Breeding Bird Survey, Canadian Breeding Bird Census, Christmas Bird Count, and banding and recovery events in the U.S. Banding Laboratory data set (without connecting banding and recovery events for individual birds at this point in the process—see below). For Mexico, observational data sets were complemented by data from the *Atlas of Mexican Bird Distributions*, which summarizes Mexican bird specimen holdings of 64 natural history museums (Peterson et al. 1998; Navarro-Sigüenza et al. 2002, 2003). For the purposes of the prototype, occurrence data were restricted to records from two 3-month "breeding" (May–July) and "wintering" periods (December–February).

Ecological niche modeling (ENM) tools were then used to interpolate between occurrence records and map likely geographic distributions for each species in each season. Specifically, I used the Genetic Algorithm for Rule-set Prediction (GARP; Stockwell and Noble 1992, Stockwell and Peters 1999) to profile each species' seasonal distribution in ecological space. Species profiles, which can be termed an "ecological niche model," can then be projected back onto the geographic landscape to identify areas of potential distribution (Peterson 2003). The approach, and its assumptions and details, is discussed at length elsewhere (Soberón and Peterson 2004, 2005).

After development of a static picture of the overall breeding and wintering distributions of each species, I used the Bird Banding Laboratory data to link areas in the two seasons. To illustrate the idea of connectivity, I selected eight 500-km-diameter "windows" across the arctic rim of North America (Fig. 6.1) as focal source regions in the prototype. My goal was to be able to anticipate where birds originating (breeding) in each of these eight windows spend the winter within the modeled winter distribution of each species under consideration.

I converted each of the banding records from the breeding season for which a winter recovery was available (or vice versa) into a vector data element in a GIS shapefile, as a line segment connecting the breeding locality and the winter locality (Fig. 6.1). For each species and for each window, I then identified vectors originating in that window and plotted a point on the winter distributional area to represent the winter destinations of those birds from that sector of the breeding distribution. I also produced 1,000 random points from across the winter distribution of each species to provide a "background" value of inferred absences of the species.

Last, assigning the known winter destinations a value of 10 and the random points values of 0, I fit a surface using an inverse distance weighting, of the form $w(d) = 1/d^p$, and used a distance weighting of $p = 2$. Weighting schemes were arbitrary and are under further experimentation for the full implementation of the forecasting system. The resulting surface provided a first approximation of the winter destinations of individuals of each species originating within each of the eight windows. The surfaces were summed across the six species for each window to provide a more general picture of the winter destinations of birds originating in those sectors of the arctic rim.

Results and Discussion

The steps in my prototype model provide information on (1) a "best guess" as to the breeding and wintering distributions of each species; and (2) to the limits of information available, the likely winter destinations of members of each species originating in particular portions of the breeding distribution. The procedure is complex, with several decisions and options, and clearly will require further exploration and fine-tuning. However, it provides a first picture of interconnectivity of areas in the breeding distribution with areas in the winter distribution, at least for migratory species that have received sufficient banding and recovery effort.

The Greater White-fronted Goose provides an illustration of the procedures for estimating connectivity between seasonal distributional areas (Fig. 6.2). The breeding range of

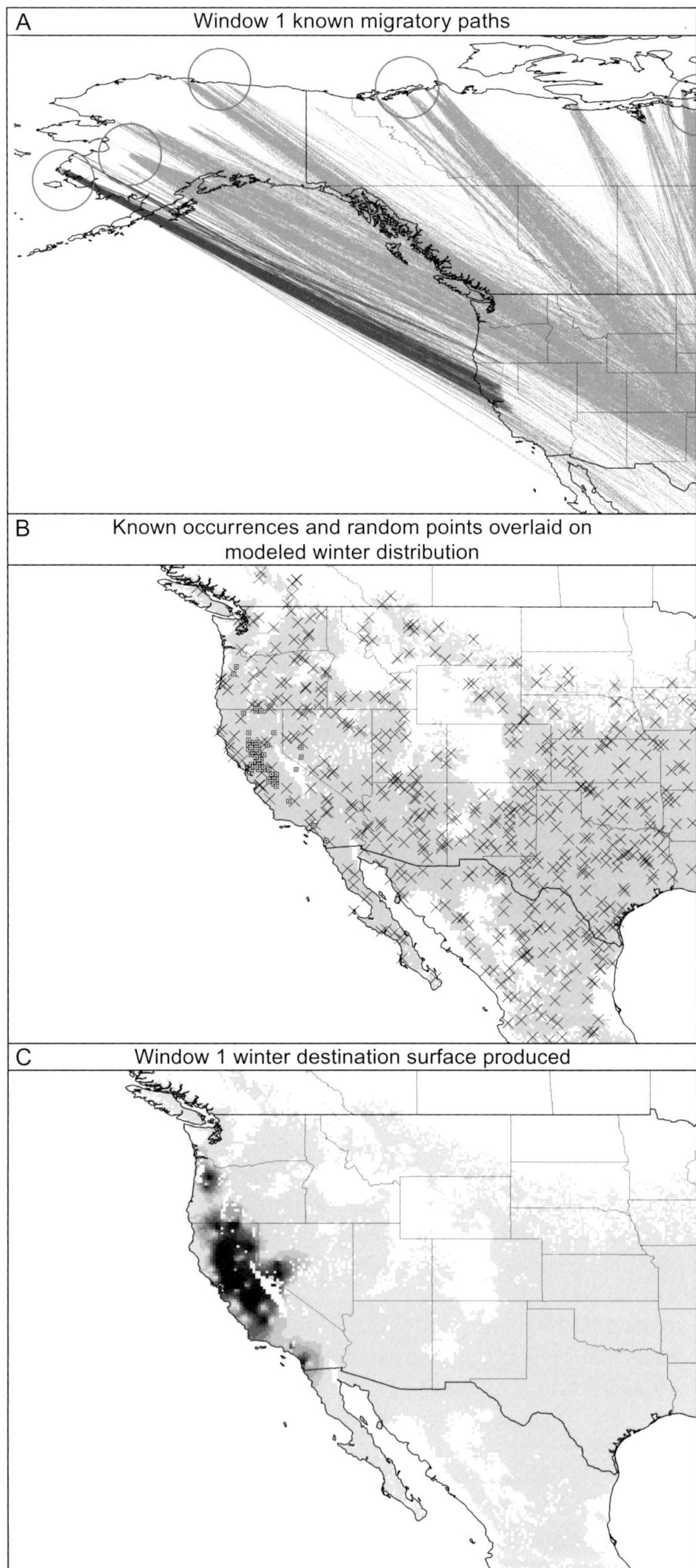

Figure 6.2. Example of Greater White-fronted Goose (*Anser albifrons*) records of breeding within Window 1, and procedures used to characterize their winter destinations. (A) Identification of paths originating in Window 1. (B) Points indicating winter destinations (dotted squares) and random points used for contrast (Xs). (C) Surface produced, in which darkest shading of gray indicates greatest numbers of individuals.

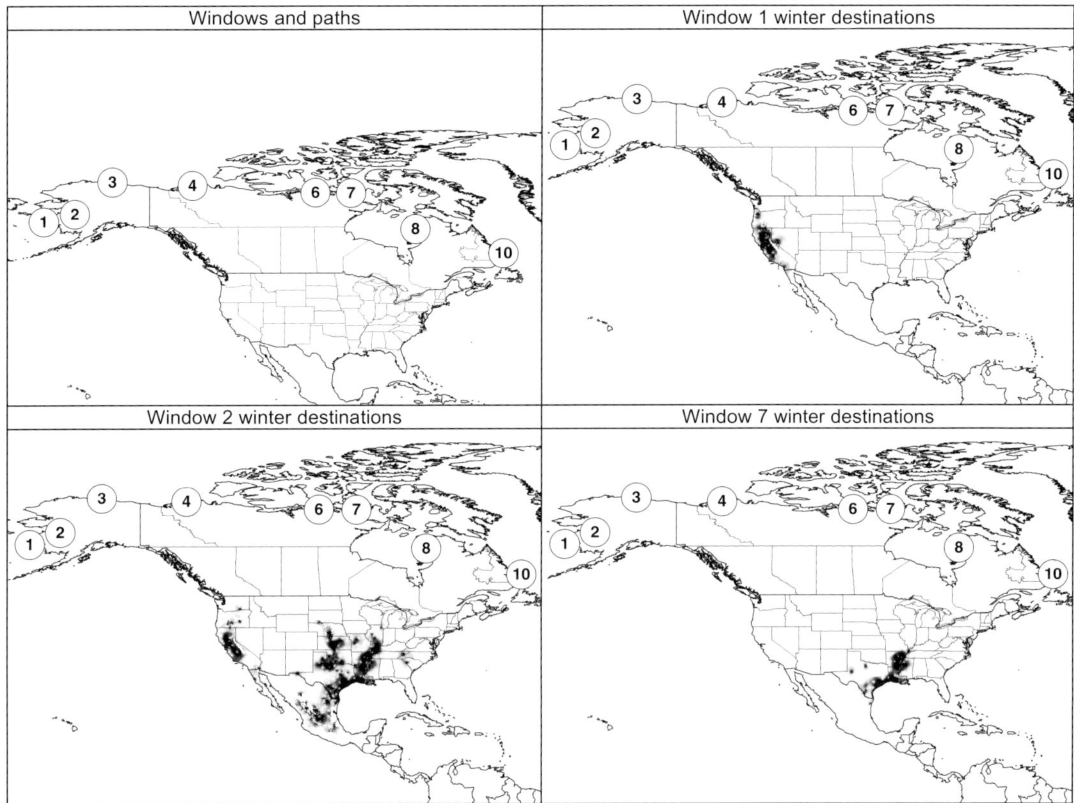

Figure 6.3. Greater White-fronted Goose (*Anser albifrons*) winter destinations, based on breeding-distribution banding records in three windows across the Arctic rim of North America.

the species extends from westernmost Alaska west to central northern Canada, and it winters in central California, the southern Great Plains, the lower Mississippi River Valley, and the Gulf Coast (Palmer 1976). As expected, but perhaps at a finer scale than would have been expected, non-random associations exist between portions of the breeding and winter distributional areas (Fig. 6.3). Birds breeding in westernmost Alaska (Window 1) migrate almost exclusively to central California; birds breeding just a few hundred kilometers farther east in western Alaska (Window 2), however, spread out between central California, the south-central United States, and central Mexico. Birds from the eastern extreme of the species' breeding distribution (Window 7), in contrast, do not migrate to California at all— rather, they winter in the lower Mississippi River Valley, along the Gulf Coast, and to a lesser degree on the East Coast. These contrasts and non-random associations are indicative of interesting

and intriguing complexities in the fine details of bird migration in North America that remain to be discovered once additional techniques (e.g., stable isotope analysis, molecular genetic studies) are developed fully for this purpose.

Looking at more general patterns by averaging over all six species reviewed in the prototype (Fig. 6.4) reveals patterns similar to those just mentioned, but the greater diversity of species makes for more general results. In particular, westernmost Alaska breeders (Window 1) concentrate in the Pacific coastal states, and again, breeders from just a few hundred kilometers inland and eastward across much of the arctic rim (Windows 2–7) spread out between central California and the interior of the United States. East of Hudson Bay (Window 8), however, the California component of wintering populations is lost entirely, and wintering is focused in the lower Mississippi River Valley and on the East Coast. Birds breeding farther to the east (Window 10) appear to winter almost exclusively along the northeastern coast of the continent, although the sample of species included in the prototype is small.

The patterns illustrated by this prototype serve several purposes—including illustration of the complexity in how breeding populations map onto wintering populations. Linking these results back into the context of HP-H5N1, avian movements indicate that the likely "next step" of the virus in North America will be highly sensitive to where the virus enters North America and to where it is able to spread before migration begins—just a few hundred kilometers could make a considerable difference in whether whole sectors of the continent should be considered vulnerable. What the prototype does not do is summarize any real patterns or offer any solid basis for policy decisions, as many more species would have to be considered before such results could be considered robust.

Prototype to Full Implementation

The prototype model has received a number of extensions and improvements. It is—for the moment—developed in terms of breeding populations mapping onto wintering populations, as this situation will likely be the first issue at hand in dealing with HP-H5N1 in North America. Clearly, extending the coverage of species to all migratory birds in North America is a key goal. Greater temporal detail is also desired, to the extent permitted by the temporal density of records available in the bird-banding data set; moving from two seasons to perhaps biweekly periods or months, at least in the months of migration, is important. Next, the prototype considered eight windows across the arctic rim of North America; an important extension is to develop it for a uniform grid across the continent (e.g., all ½° or 1° grid squares in North America). Last, several technical details of the analysis need attention: seasonal distributional models should be tested and evaluated for significant predictivity prior to use, more diverse surface-fitting algorithms should be explored, and a number of sample size–dependent options should be added to the analysis.

With additional model improvements—which will, of course, require significant investment of time and computing—the prototype can cease to be a prototype and begin to be a summary of broadest-scale information resources regarding the details of the geography of North American bird migration. One of the principal lessons of this exercise will be that of illustrating the combination of great complexity with sparse data and knowledge—much more remains to be learned about bird migration, even in North America (Palmer 1976). A knowledge base on avian movements will be useful in anticipating likely patterns of spread once HP-H5N1 reaches North America.

RECOMMENDATIONS AND CONCLUSIONS

The appearance of HP-H5N1 and the potential risk of influenza pandemics represent a serious challenge to a broad sector of the scientific community—obviously, the public health and epidemiological communities must worry about transmission among humans and the possibility of an emerging influenza pandemic. However, HP-H5N1 as an avian zoonosis also presents challenges to ornithologists worldwide—understanding the fine details of migratory bird movements to permit anticipating of next steps in the spread of the virus globally. Ornithology has been a scientific discipline for several hundred years and should be able to inform the public health and agriculture sectors about how birds will likely interact with this virus. Many lessons remain, however, particularly in the realm of organizing, sharing, and interpreting information already in hand.

Several sectors of this challenge would benefit from increased attention from the ornithological community. In particular, of particular need is intensive monitoring of bird movements within continents during migratory periods; such data would greatly inform efforts to anticipate when a disease like HP-H5N1 would arrive at a particular location, given a previous appearance elsewhere. Spatial data are now being organized in data caches such as ORNIS (ornisnet. org, mostly for specimen-based data) and eBird (ebird.org/content/ebird, for observation-based data). In particular, focus on documenting frequencies and patterns of intercontinental avian movements would be helpful—this information will require intensive documentation of occurrences and movements in remote areas of western Alaska, northeastern Canada, and Greenland, as well as in Australia, New Guinea, and South America. More generally, broad, efficient, and open sharing of distributional data of all sorts

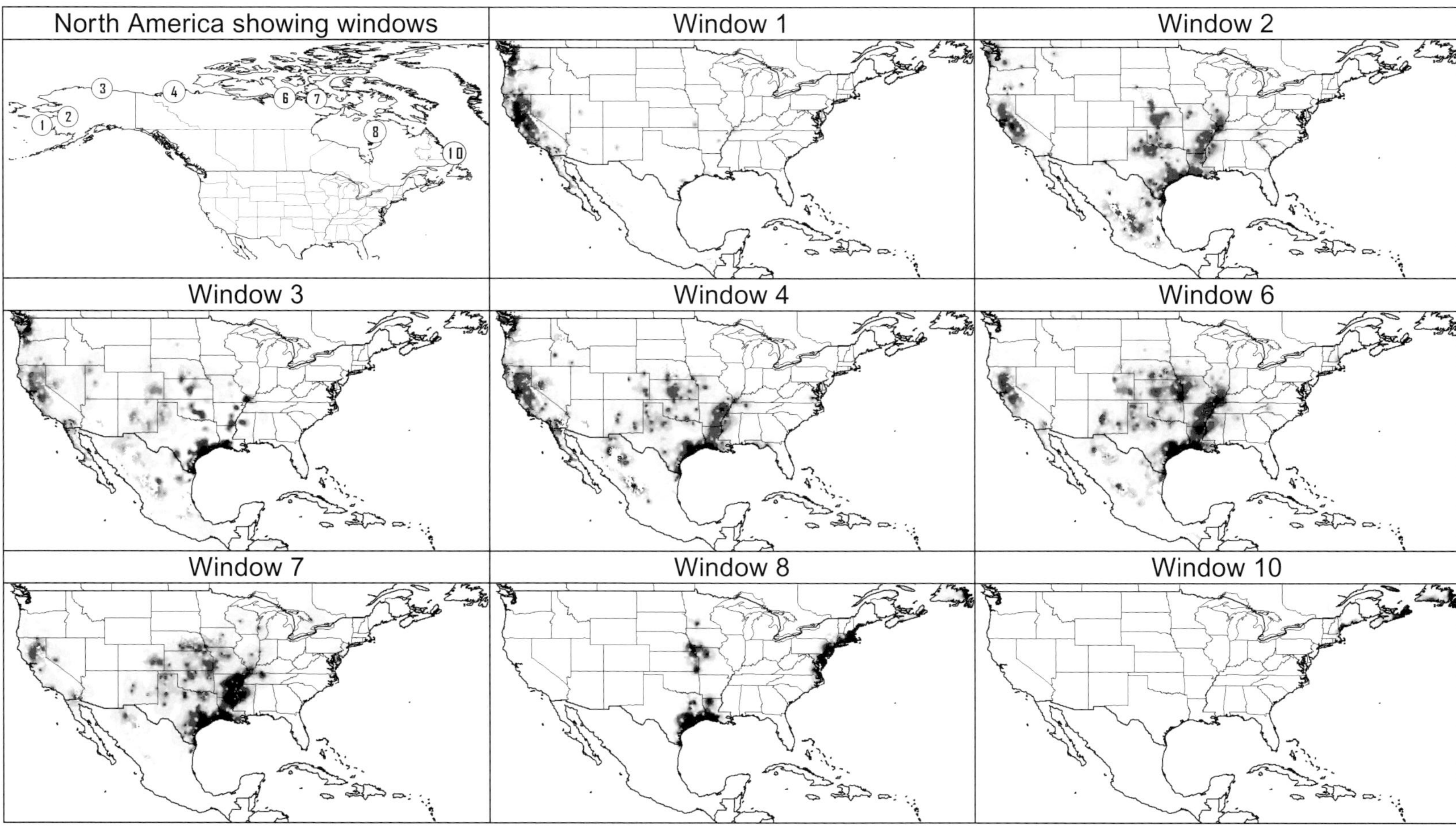

Figure 6.4. General patterns of winter destinations across the six species of waterfowl included in the Arctic rim prototype, showing estimated winter destinations of individuals breeding within eight windows across the region.

via new Internet-based protocols such as DiGIR (Stein and Wieczorek 2004) will greatly enable such research (Peterson et al. 2005).

Further development and exploration of genetic and isotopic approaches to understanding the fine details of migration of key species will also be important in coming years (Webster et al. 2002). These techniques have the potential to be more direct and less subject to sampling bias than the banding data used herein—at least once they are developed and tested more rigorously. New methods offer the potential for creating connectivity maps between breeding and wintering areas for species that do not depend on the vagaries of banding data, and would depend simply on the existence of robust protocols for analysis of recent, data-rich specimen material in natural history museum collections.

Last, management of HP-H5N1 will certainly require major efforts in the realm of actual monitoring for the virus. Such work is likely to be perilous, as the virus is highly pathogenic to humans and any handling of potentially infected birds must be carried out under the strictest of safety guidelines. However, it will track the advance of the virus, and is the only means by which to clarify the points discussed above regarding wild birds and landbirds as contributors to its spread. Virus tracking would involve sampling birds and vouchering the samples with specimens, sampling live birds to boost numbers and increase probabilities of detection of rare infections, and detecting dead birds and getting them tested. Screening of dead birds proved to be among the most sensitive "detectors" of the advance of West Nile virus across North America in the past half decade (Eidson et al. 2001).

ACKNOWLEDGMENTS

I thank M. Ruggiero, G. Cotter, J. Hill, and others at the National Biological Information Infrastructure for their assistance in obtaining data critical to the development of the prototype, and S. Kelling, T. B. Smith, and K. Winker for useful discussions of the topic. M. Robbins provided helpful comments on an early draft of the manuscript.

LITERATURE CITED

ABC. 2005. Avian influenza spreads: Role of wild birds debated. American Bird Conservancy. <http://www.abcbirds.org/media/releases/avian_influenza.htm>.

Alexander, D. J. 2007. Summary of avian influenza activity in Europe, Asia, Africa, and Australasia, 2002–2006. Avian Diseases 51:161–166.

Atkinson, P., A. Baker, R. Bevan, N. Clark, K. Cole, P. Gonzalez, J. Newton, L. Niles, and R. Robinson. 2005. Unravelling the migration and moult strategies of a long-distance migrant using stable isotopes: Red Knot *Calidris canutus* movements in the Americas. Ibis 147:738–749.

Belshe, R. B. 2005. The origins of pandemic influenza—lessons from the 1918 virus. New England Journal of Medicine 353:2209–2211.

Berthold, P., E. Nowak, and U. Querner. 1995. Satellite-tracking of a migratory bird from Central-Europe to South-African winter quarters: a case-report of the White Stork. Journal für Ornithologie 136:73–76.

BirdLife International. 2006. BirdLife statement on avian influenza. BirdLife International. <www.birdlife.org/action/science/species/avian_flu/>.

Brockmann, D., L. Hufnagel, and T. Geisel. 2006. The scaling laws of human travel. Nature 439:462–466.

Chen, H., G. J. D. Smith, K. S. Li, J. Wang, X. F. Fan, J. M. Rayner, D. Vijaykrishna, J. X. Zhang, L. J. Zhang, C. T. Guo, C. L. Cheung, K. M. Xu, L. Duan, K. Huang, K. Qin, Y. H. C. Leung, W. L. Wu, H. R. Lu, Y. Chen, N. S. Xia, T. S. P. Naipospos, K. Y. Yuen, S. S. Hassan, S. Bahri, T. D. Nguyen, R. G. Webster, J. S. M. Peiris, and Y. Guan. 2006. Establishment of multiple sublineagues of H5N1 influenza virus in Asia: implications for pandemic control. Proceedings of the National Academy of Sciences USA 103:2845–2850.

Cramp, S., and C. M. Perrins (editors). 1977–1994. The birds of the Western Palearctic: handbook of the birds of Europe, the Middle East, and North Africa. Oxford University Press, Oxford, UK.

Dupuis, A. P., 2nd, P. P. Marra, and L. D. Kramer. 2003. Serologic evidence of West Nile virus transmission, Jamaica, West Indies. Emerging Infectious Diseases 9:860–863.

Dupuis, A. P., 2nd, P. P. Marra, R. Reitsma, M. J. Jones, K. L. Louie, and L. D. Kramer. 2005. Serologic evidence for West Nile virus transmission in Puerto Rico and Cuba. American Journal of Tropical Medicine and Hygiene 73:474–476.

Eidson, M., L. D. Kramer, W. Stone, Y. Hagiwara, K. Schmitt, and N.S.W.N.V.A.S. Team. 2001. Dead bird surveillance as an early warning system for West Nile virus. Emerging Infectious Diseases 7:631–635.

FAO. 2005. Update on the avian influenza situation (as of 01/09/2005)—issue no. 33. FAOAIDEnews, FAO, Rome, Italy.

Farmer, A., M. Abril, M. Fernandez, J. Torres, C. Kester, and C. Bern. 2004. Using stable isotopes to associate migratory shorebirds with their wintering locations in Argentina. Ornitologia Neotropical 15:377–384.

Hemachudha, T., J. Laothamatas, and C. E. Rupprecht. 2002. Human rabies: a disease of complex neuropathogenetic mechanisms and diagnostic challenges. Lancet Neurology 1:101–109.

Hobson, K. 1999. Stable-carbon and nitrogen isotope ratios of songbird feathers grown in two terrestrial biomes: implications for evaluating trophic relationships and breeding origins. Condor 101:799–805.

Hobson, K., E. Sinclair, A. York, J. Thomason, and R. Merrick. 2004. Retrospective isotopic analyses of Steller sea lion tooth annuli and seabird feathers: a cross-taxa approach to investigating regime and dietary shifts in the Gulf of Alaska. Marine Mammal Science 20:621–638.

Hsieh, Y. C., T. Z. Wu, D. P. Liu, P. L. Shao, L. Y. Chang, C. Y. Lu, C. Y. Lee, F. Y. Huang, and L. M. Huang. 2006. Influenza pandemics: past, present and future. Journal of the Formosan Medical Association 105:1–6.

Hulse-Post, D. J., K. M. Sturm-Ramirez, J. Humberd, P. Seiler, E. A. Govorkova, S. Krauss, C. Scholtissek, P. Puthavathana, C. Buranathai, T. D. Nguyen, H. T. Long, T. S. Naipospos, H. Chen, T. M. Ellis, Y. Guan, J. S. Peiris, and R. G. Webster. 2005. Role of domestic ducks in the propagation and biological evolution of highly pathogenic H5N1 influenza viruses in Asia. Proceedings of the National Academy of Sciences USA 102:10682–10687.

JNCC. 2006. Position statement on avian influenza. Joint Nature Conservation Committee. <http://www.jncc.gov.uk/page-3519>.

Kilpatrick, A. M., A. A. Chmura, D. W. Gibbons, R. C. Fleischer, P. P. Marra, and P. Daszak. 2006. Predicting the global spread of H5N1 avian influenza. Proceedings of the National Academy of Sciences USA 103:19368–19373.

Kimura, M., S. M. Clegg, I. J. Lovette, K. R. Holder, D. J. Girman, B. Mila, P. Wade, and T. B. Smith. 2002. Phylogeographical approaches to assessing demographic connectivity between breeding and overwintering regions in a nearctic-neotropical warbler (*Wilsonia pusilla*). Molecular Ecology 11:1605–1616.

Komar, N. 2003. West Nile virus: epidemiology and ecology in North America. Advances in Virus Research 61:185–234.

Komar, O., M. B. Robbins, G. Guzmán-Contreras, B. W. Benz, K. Klenk, B. J. Blitvich, N. L. Marlenee, K. L. Burkhalter, S. Beckett, G. Gonzálvez, C. J. Peña, A. T. Peterson, and N. Komar. 2005. West Nile virus survey of birds and mosquitoes in the Dominican Republic. Vector-Borne and Zoonotic Diseases 5:120–126.

Komar, O., M. B. Robbins, K. Klenk, B. J. Blitvich, N. L. Marlenee, K. L. Burkhalter, D. J. Gubler, G. Gonzálvez, C. J. Peña, A. T. Peterson, and N. Komar. 2003. West Nile virus transmission in resident birds, Dominican Republic. Emerging Infectious Diseases 9:1299–1302.

Kou, Z., F. M. Lei, J. Yu, Z. J. Fan, Z. H. Yin, C. X. Jia, K. J. Xiong, Y. H. Sun, X. W. Zhang, X. M. Wu, X. B. Gao, and T. X. Li. 2005. New genotype of avian influenza H5N1 viruses isolated from tree sparrows in China. Journal of Virology 79:15460–15466.

Kwon, Y. K., H. W. Sung, S. J. Joh, Y. J. Lee, M. C. Kim, J. G. Choi, E. K. Lee, S. H. Wee, and J. H. Kim. 2005. An outbreak of highly pathogenic avian influenza subtype H5N1 in broiler breeders, Korea. Journal of Veterinary and Medical Science 67:1193–1196.

Lavanchy, D. 1998. The WHO update on influenza A (H5N1) in Hong Kong. European Surveillance 3:23–25.

Lott, C., T. Meehan, and J. Heath. 2003. Estimating the latitudinal origins of migratory birds using hydrogen and sulfur stable isotopes in feathers: influence of marine prey base. Oecologia 134:505–510.

Marra, P. P., S. M. Griffing, and R. G. McLean. 2003. West Nile virus and wildlife health. Emerging Infectious Diseases 9:898–899.

Melville, D. S., and K. F. Shortridge. 2006. Spread of H5N1 avian influenza virus: an ecological conundrum. Letters in Applied Microbiology 42:435–437.

Mermel, L. A. 2005. Pandemic avian influenza. Lancet Infectious Diseases 5:666–667.

Murphy, F. A., M. P. Kiley, and S. P. Fisher-Hoch. 1990. Filoviridae: Marburg and Ebola viruses. Pp. 933–942 *in* B. N. Fields and D. M. Knipe (editors), Virology. Raven Press, Ltd., New York, NY.

National Audubon Society. 2005. Avian influenza information. National Audubon Society. <http://www.audubon.org/bird/avianflu/avianflu.htm>.

Navarro-Sigüenza, A. G., A. T. Peterson, and A. Gordillo-Martínez. 2002. A Mexican case study on a centralised database from world natural history museums. CODATA Journal 1:45–53.

Navarro-Sigüenza, A. G., A. T. Peterson, and A. Gordillo-Martínez. 2003. Museums working together: the atlas of the birds of Mexico. Bulletin of the British Ornithologists' Club 123A:207–225.

Normile, D. 2005. Avian influenza: Europe scrambles to control deadly H5N1 strain. Science 310:417.

Olsen, B., V. J. Munster, A. Wallensten, J. Waldenström, A. D. M. E. Osterhause, and R. A. M. Fouchier. 2006. Global patterns of Influenza A virus in wild birds. Science 312:384–388.

Palmer, R. S. (editor). 1976. Handbook of North American birds, Vol. 2: Waterfowl (first part). Yale University Press, New Haven, CT.

Parry, J. 2006. Children in Turkey die after contracting avian flu. British Medical Journal 332:67.

Peterson, A. T. 2003. Predicting the geography of species' invasions via ecological niche modeling. Quarterly Review of Biology 78:419–433.

Peterson, A. T., M. J. Andersen, S. Bodbyl-Roels, P. Hosner, Á. Nyári, C. Oliveros, and M. Papeş period 2009. A prototype forecasting system for bird-borne disease spread in North America based on migratory bird movements. Epidemics 1:240–249.

Peterson, A. T., B. W. Benz, and M. Papeş. 2007. Highly pathogenic H5N1 avian influenza: entry pathways into North America via bird migration. PLoS ONE 2:e261.

Peterson, A. T., C. Cicero, and J. Wieczorek. 2005. Free and open access to distributed data from bird specimens: why? The Auk 122:987–990.

Peterson, A. T., N. Komar, O. Komar, A. G. Navarro-Sigüenza, M. B. Robbins, and E. Martínez-Meyer. 2004. West Nile virus in the New World: potential impacts on bird species. Bird Conservation International 14:215–232.

Peterson, A. T., A. G. Navarro-Sigüenza, and H. Benítez-Díaz. 1998. The need for continued scientific collecting: a geographic analysis of Mexican bird specimens. Ibis 140:288–294.

Soberón, J., and A. T. Peterson. 2004. Biodiversity informatics: managing and applying primary biodiversity data. Philosophical Transactions of the Royal Society of London Series B 359:689–698.

Soberón, J., and A. T. Peterson. 2005. Interpretation of models of fundamental ecological niches and species' distributional areas. Biodiversity Informatics 2:1–10.

Stein, B. R., and J. Wieczorek. 2004. Mammals of the world: MaNIS as an example of data integration in a distributed network environment. Biodiversity Informatics 1:14–22.

Stockwell, D. R. B., and I. R. Noble. 1992. Induction of sets of rules from animal distribution data: a robust and informative method of analysis. Mathematics and Computers in Simulation 33:385–390.

Stockwell, D. R. B., and D. P. Peters. 1999. The GARP modelling system: problems and solutions to automated spatial prediction. International Journal of Geographical Information Science 13:143–158.

Taubenberger, J. K., and D. M. Morens. 2006. Influenza revisited. Emerging Infectious Diseases 12:1–2.

Tran, T. H., T. L. Nguyen, T. D. Nguyen, T. S. Luong, P. M. Pham, V. C. Nguyen, T. S. Pham, C. D. Vo, T. Q. Le, T. T. Ngo, B. K. Dao, P. P. Le, T. T. Nguyen, T. L. Hoang, V. T. Cao, T. G. Le, D. T. Nguyen, H. N. Le, K. T. Nguyen, H. S. Le, V. T. Le, D. Christiane, T. T. Tran, J. de Menno, C. Schultsz, P. Cheng, W. Lim, P. Horby, and J. Farrar. 2004. Avian influenza A (H5N1) in 10 patients in Vietnam. New England Journal of Medicine 350:1179–1188.

Webster, M. S., P. P. Marra, S. M. Haig, S. Bensch, and R. T. Holmes. 2002. Links between worlds: unraveling migratory connectivity. Trends in Ecology & Evolution 17:76–83.

Webster, R. G., M. Peiris, H. Chen, and Y. Guan. 2006. H5N1 outbreaks and enzootic influenza. Emerging Infectious Diseases 12:3–8.

Winker, K., K. G. McCracken, D. D. Gibson, C. L. Pruett, R. Meier, F. Huettmann, M. Wege, I. V. Kulikova, Y. N. Zhuravlev, M. L. Perdue, E. Spackman, D. L. Suarez, and D. E. Swayne. 2007. Movements of birds and avian influenza from Asia into Alaska. Emerging Infectious Diseases 13:547–552.

World Health Organization. 2005. H5N1 avian influenza: timeline. World Health Organization. <http://www.who.int/csr/disease/avian_influenza/Timeline_28_10a.pdf>.

Yuanji, G. 2002. Influenza activity in China: 1998–1999. Vaccine 20(Supplement 2):S28–35.

CHAPTER SEVEN

Immunophenotyping of Avian Lymphocytes

IMPLICATIONS AND FUTURE FOR UNDERSTANDING DISEASE IN BIRDS

*Jeanne M. Fair, Kirsten J. Taylor-McCabe,
Yulin Shou, and Babetta L. Marrone*

Abstract. Chicken (*Gallus gallus domesticus*) T-cell populations can be delineated into subsets based on their expression of cell-surface proteins such as cluster of differentiation (CD) cell surface markers. However, immunophenotyping using flow cytometry in birds has focused on cell characterization in the thymus and spleen during development in chickens. West Nile virus (WNV) causes differential infections in birds, ranging the entire spectrum of pathogenesis. In order to accurately assess immunocompetence to diseases such as WNV in birds, more efficient methodology to access natural variability in avian immune function must be devised and understood. Previously, lymphocyte subpopulations CD4$^+$ and CD8$^+$ have been found to be critical for clearing infection of WNV in mammals. Focusing on chickens, a species that is susceptible but not infective for WNV, our objectives were to: (1) further develop flow cytometry for estimating subpopulations of lymphocytes in peripheral blood from poultry, (2) estimate the best antibody and cell marker combination for estimating lymphocyte subpopulations, and (3) estimate repeatability and application to other avian species susceptible to WNV. Immunophenotyping of CD3+, CD4+, CD8+, and CD45+ was successfully completed for chicken peripheral blood but not for the Common Raven (*Corvus corax*) or Black-billed Magpie (*Pica hudsonia*). Future studies include immunophenotyping during infection studies of WNV in chickens and further development of flow cytometry for other bird species.

Key Words: chicken, flow cytometry, *Gallus gallus domesticus*, host range, immunophenotyping, lymphocytes, West Nile virus.

Caracterización Inmunofenotípica de Linfocitos de Aves: Las Implicaciones y el Futuro Para el Entendimiento de la Enfermedad en Las Aves

Resumen. Las poblaciones de las células T del pollo (*Gallus gallus domesticus*) pueden ser caracterizadas en base a la expresión celular de sus proteínas de superficie, tales como los marcadores de antígenos de diferenciación leucocitaria (CD). Sin embargo, la caracterización inmunofenotípica usando citometría de flujo en aves se ha enfocado a la caracterización celular en el timo y el bazo durante el desarrollo de pollos. El virus del Nilo Occidental (VNO) causa diferentes formas de infección en las aves, y cubre todo el espectro patológico. Para poder evaluar de manera precisa la inmunocompetencia de las aves a las enfermedades como el VNO es necesario desarrollar metodologías más eficientes, que permitan accesar la variabilidad natural de la función

Fair, J. M., K. J. Taylor-McCabe, Y. Shou, and B. L. Marrone. 2012. Immunophenotyping of avian lymphocytes: implications and future for understanding disease in birds. Pp. 81–90 *in* E. Paul (editor). Emerging avian disease. Studies in Avian Biology (vol. 42), University of California Press, Berkeley, CA.

inmunológica de las aves. Anteriormente se identificó que las subpoblaciones de linfocitos CD4+ y CD8+ son indispensables para eliminar la infección del VNO en los mamíferos. Enfocándonos en los pollos, la cuál es una especie susceptible pero no infectiva para el VNO, nuestros objetivos fueron: (1) mejorar la metodología de la citometría de flujo para estimar subpoblaciones de linfocitos en la sangre periférica de aves de corral, (2) calcular la mejor combinación de anticuerpos y marcadores celulares para estimar subpoblaciones de linfocitos, y (3) estimar la consistencia y la aplicación para otras especies de aves susceptibles al VNO. La caracterización inmunofenotípica de CD3+, CD4+, CD8+ y CD45+ fue exitosa para la sangre periférica de pollos, pero no fue exitosa para el Cuervo Común (*Corvus corax*) y para la Urraca (*Pica hudsonia*). Los estudios para el futuro deberán incluír la caracterización inmunofenotípica de los pollos durante la infección con el VNO y el desarrollo de la citometría de flujo para otras especies de aves.

Palabras Clave: caracterización inmunofenotípica, citometría de flujo, *Gallus gallus domesticus*, linfocitos, pollo, rango de hospedero, virus del Nilo Occidental.

R ecently, considerable interest has focused on life-history effects on immunocompetence and emerging diseases in birds. If immunocompetence is limited by available resources, then trade-offs between investment in life-history components and investment in immunocompetence could be important in determining optimal life-history traits (Norris and Evans 2000). Understanding these trade-offs and immunocompetence in respect to emerging avian diseases will be critical for predicting, mitigating, and managing disease impacts in both domestic and wild bird populations.

Immunophenotyping can be defined as a process used to identify cells based on the types of antigens or markers on the surface of the cell. Specific antibodies are the critical tools for immunophenotyping, and antibodies can be colored by conjugation with the appropriate fluorochrome used in flow cytometry (Stewart and Nicholson 2000). Antibodies are globular proteins known as immunoglobulins (Ig; Kindt et al. 2007). Since great biological diversity exists among proteins of the same types within and between species, the differences can be used to elicit the production of specific antibodies. However, these differences typically result in non-cross-reactivity between antibodies produced by other species that are not phylogenetically related. For example, the antibodies for a subpopulation of lymphocyte T cells in humans may cross-react with T cells from monkeys as the T cells share the common epitope (Stewart and Nicholson 2000). Immunophenotyping is accomplished by identifying cells using fluorochrome-conjugated antibodies as probes to proteins expressed by cells. Lymphocytes express distinct assortments of molecules on their cell surfaces, many of which reflect either different stages of their lineage-specific differentiation or different states of activation or inactivation. Lymphocyte cell surface molecules are routinely detected with anti-leukocyte monoclonal antibodies (mAbs). Using different combinations of mAbs, it is possible to profile the cell surface immunophenotypes of different leukocyte subpopulations, including the functionally distinct mature lymphocyte subpopulations of B cells, helper T cells, cytotoxic T cells, and natural killer cells, as well as estimate the whole leukocyte count. The cluster of differentiation (CD) applies to the subpopulations of lymphocytes that describe the antibody directed to the antigen on the cell, and not to an epitope of that antigen (Table 7.1). For example, in humans, Stewart and Nicholson (2000) noted that CD11a refers to all antibodies that bind to any epitopes on the alpha chain of LFA1. Conventions in nomenclature arose out of the need to identify the antigen, rather than the epitope.

The CD molecules most commonly referred to are CD4+ and CD8+, which are markers for two different subtypes of T-lymphocytes, T-helper cells, and cytotoxic T cells, respectively, with different roles in the immune system, as their names imply. Lymphocytes that express CD4+ and CD8+ are critical for the regulation and control of many pathogens—in particular, viral pathogens. For example, CD4+ is specifically recognized and bound by several viruses such as HIV, leading to viral infection and destruction of

Known specificity and function of monoclonal antibodies for antigens in mammals.

Cell surface antigens	Specificity/distribution	Conjugation/ label used	Function
CD3+	T cytotoxic lymphocytes, during thympoiesis	FITC, PE	T-cell receptors
CD4+	T helper lymphocytes, some natural killer (NK) cells	FITC, PE	Antigen recognition, MHC Class I and II
CD8+	T cytotoxic lymphocytes, thymocytes, some NK cells	FITC, PE	Antigen recognition, MHC Class I restricted
CD45+	Total lymphocytes, all nucleated haematopoietic cells	FITC, PE	Regulation of tyrosine phosporylation, signaling thresholds

CD4+ T cells (Altfeld and Rosenberg 2000, Picker and Maino 2000). Evidence also suggests that CD4+ and CD8+ are important in the immunology of West Nile virus (WNV) in vertebrate hosts (Wang et al. 2003b, Shrestha and Diamond 2004, Shrestha et al. 2006, Sitati and Diamond 2006). Recent techniques have improved our ability to quantify CD4+ and CD8+ T-cell responses following viral infection. It is becoming clearer that many viral infections induce strong antigen-specific T-cell responses (Doherty and Christensen 2000). Due to the importance of their response to disease in vertebrates, T cells are, likewise, considered a critical aspect of the immune system to investigate in birds. Flow cytometry rarely has been applied to birds compared to humans, and while most immunophenotyping using flow cytometry has focused on cell characterization in the thymus and spleen and during the development in chickens (*Gallus gallus domesticus*; Lillehoj 1991, Paramithiotis et al. 1991, Erf et al. 1998), few studies have immunophenotyped peripheral blood in chickens (Furusawa et al. 2000, Kliger et al. 2000, Czekaj et al. 2005, Bohls et al. 2006).

Immunophenotyping Differences Between Birds and Humans   Birds differ from humans in that they have nucleated red blood cells that allow for cell division. In the past, this difference in physiology has provided a roadblock to rapidly and accurately obtaining lymphocyte cell counts in birds since red cells outnumber white cells by

orders of magnitude. For human lymphocyte cell measurements, non-nucleated red blood cells are easily lysed and can be eliminated or separated from measurement. However, avian red blood cells are not as easily lysed without lysing all the lymphocytes simultaneously, and red blood cells also been found to have a lower ability to adapt to local sheer force such as centrifugation (Gaehtgens et al. 1981).

Therefore, centrifuging to separate lymphocytes can increase the amount of erythrocytic lysate in the sample, increasing the need for multiple cleansings. Another major challenge for sampling birds is the amount of sample that can be obtained nondestructively. By using the assay platform of single-cell, high-sensitivity flow cytometry, small sample volume can be overcome. Because thousands of cells per second can be analyzed individually, the counting statistics are favorable for examining scant sample volumes. In addition, flow cytometry is especially well suited for high-throughput analysis of large numbers of samples. In this paper, we used a two-color labeling system because only FITC, unlabeled, or R-PE labels are available for anti-chicken CD markers. We are confident that once this technique is utilized more often in avian samples, assays for more than two colors, such as those used commonly for human studies, will become available. Use of the technique will enable more immune markers to be processed within one sample within a given period of time than has been done previously, allowing for immediate analysis. Collecting samples from birds either in the laboratory or in the field is very time consuming and limits the number of samples. The ability to sample more birds at greater frequencies will overcome another limitation of the use of free-range avian sentinels, which is the potential for grossly underestimating the effect of disease because of poor sampling.

Immunology of an Example Disease: West Nile Virus Since the emergence of WNV in New York City in 1999, this *Flavivirus* has emerged as one of the most important arthropod-borne viruses in North America (for a review of WNV in birds, see McLean 2006). WNV has been maintained in an enzootic cycle among mosquitoes, birds and mammals, and the environment that affects mosquito populations. One of the most striking impacts of WNV was the difference in host susceptibility among species of birds, ranging from totally asymptomatic species that may still replicate the virus, such the Rock Pigeon (*Columba livia*; Allison et al. 2004, Deegan et al. 2005, Gibbs et al. 2005), to species that suffer extremely high morality rates such as the Common Raven (*Corvus corax*; Potter 2004) and the American Crow (*C. brachyrhynchos*; Yaremych et al. 2004, Caffrey et al. 2005, Ward et al. 2006). Komar et al. (2003) experimentally showed five top species that might have a higher probability of functioning as reservoir species based on susceptibility, infectiousness, and duration of viral shedding. These species included the Blue Jay (*Cyanocitta cristata*), Common Grackle (*Quiscalus quiscula*), House Finch (*Carpodacus mexicanus*), American Crow, and House Sparrow (*Passer domesticus*). Another long-term study using Breeding Bird Survey data from the last three decades has shown population decreases as high as 45% in groups of birds such as crows and ravens since 2001 across the U.S. (LaDeau et al. 2007). Currently, the mechanisms underlying differences in bird species' susceptibilities remain unknown for WNV.

The host range of pathogens can be driven by many things. The major host range determinant for all influenzas is the binding site for entrance of the virus into the cell, sialic acid, which varies between mammals and bird cell membranes (Suzuki et al. 2000). To gain entrance into cells, human influenza viruses bind preferentially to sialic acid containing N-acetylneuraminic acid alpha2,6-galactose (SAalpha2,6Gal) linkages, while avian and equine viruses bind preferentially to the sialic acid containing N-acetylneuraminic

acid alpha2,3-galactose (SAalpha2,3Gal) linkages. Another candidate host cell receptor for WNV is thought to be αVβ3 integrin (Chu and Ng 2004). Integrins are membrane proteins that play a role in the attachment of the cell to the extracellular matrix and in signal transduction. However, little information is available on integrins in birds.

In humans, host factors in elderly and immuno-compromised individuals determine a greater risk for WNV infection (Lanciotti et al. 1999). Similarly, wild birds may be immunocompromised due to environmental stress conditions and infection with other disease, leading to differential infection by WNV. Through animal models, T and B cell lymphocytes have been shown to protect against WNV infection (Halevy et al. 1994; Diamond et al. 2003a, 2003b). Humoral immunity mediated by B cells has also been shown to be a critical component of the immune response to WNV for both IgG (Diamond et al. 2003a) and IgM (Diamond et al. 2003b).

CD4+ T cells contribute to the control of WNV infection through the multiple mechanisms that include CD8+ T-cell priming, cytokine production, B-cell activation and priming, and direct cytotoxicity (Samuel and Diamond 2006). Mice deficient in T-cell receptor ß (TCRß) have increased mortality to WNV (Wang et al. 2003a). In these mice, CD4+ cells respond most in the peripheral blood (Kulkarni et al. 1991), and CD8+ cells are found in the spleen and brain following WNV infection (Liu et al. 1989).

The γδT cells are a subset of T cells that comprise a minority of CD3+ cells in the lymphoid tissue of mammals, as well as in the epithelial and mucosal sites, but are not well represented in the peripheral blood (Hayday 2000), and it has been shown that γδT CD3+ help control WNV in mice (Wang et al. 2003a). CD45+ is expressed on B lymphocytes and naïve, activated, and memory T lymphocytes.

Similar to humans where pathogenesis of WNV infection is a balance among virulence, innate and adaptive immunity, and viral evasion (Samuel and Diamond 2006), wild birds have the additional impacts of the environment and potential natural selection through past mortalities that may cause more pronounced roles over similar human situations. Another host factor suggested for susceptibility to WNV was body temperature, but it has been found that the North American genotype (NY99) has the ability to replicate at high temperatures measured in infected American

Crows (Kinney et al. 2006). While many host factors may be involved in WNV replication within individuals (Brinton 2001), differences between species should be more pronounced.

Chickens have been key players in the understanding and potential control of WNV. A candidate WNV surveillance sentinel species such as the chicken would be susceptible to mosquito-borne infection yet resistant to disease. The sentinel species must survive infection and develop detectable antibodies, but should not develop sufficient viremia to infect mosquitoes and other nearby chickens (Langevin et al. 2001). In addition, chickens have been shown to have passive transfer of maternal WNV antibodies to chicks that decayed by 28 d post-hatch without vertical transmission (Nemeth and Bowen 2007).

In order to accurately assess immunocompetence in birds, natural variability in both avian immune function and the methodology must be better measured and understood. It is also important to develop a better understanding of the pathology of disease in birds, and the host range of viruses such as WNV in birds is needed. Prior to understanding how lymphocyte subpopulations control WNV infection in different species with varying susceptibilities, immunophenotyping techniques must be further developed.

The first objective of this research was to estimate peripheral blood lymphocytes CD3+, CD4+, CD8+, and CD45+ using flow cytometry for chickens under free-ranging conditions. The second objective was to investigate repeatability of using flow cytometry for lymphocyte subpopulations and to estimate variation in a diverse selection of chicken breeds in free-ranging conditions. The last objective was to explore the use of antibody and T-cell marker combinations for estimating lymphocyte subpopulations in other bird species that are susceptible to WNV but that are not related to chickens.

METHODS

Eighty-five female chickens were obtained from several flocks maintained in and around northern New Mexico. Chickens in this study were free-ranging in order to adequately represent variability in diet and environments. Variability in the diet and environment came from different farms and husbandry practices, including diet, which was unknown. All chickens included were considered to be American breed, single-comb leghorns, either White Plymouth Rocks or New Hampshire Red varieties. Whole blood samples were collected from the brachial wing vein (ca. 2 ml) using heparinized syringes and transferred into EDTA tubes. Processing of blood samples in the laboratory began within two hours of collection. Blood was also collected from ten Black-billed Magpies (*Pica hudsonia*) and eight Common Ravens captured with the Netlauncher™ (Coda Enterprises Inc., Mesa, AZ) in Santa Fe and Los Alamos counties, respectively. Rock Pigeons captured in pigeon cage traps were also tested for immunophenotyping with anti-chicken antibodies ($n = 70$).

Lymphocyte Isolation

Venous blood samples were diluted 1:1 with sterile PBS (1X; Gibco, Carlsbad, CA) with 1% BSA (ICN Biomedicals, Aurora, OH) to a 2-ml total volume. This mixture was then layered on 3 ml of Ficoll (Histopaque-1077; Sigma, St. Louis, MO) in a 15-ml tube. The samples were centrifuged using a Centra GP8 centrifuge (IEC, Needham Heights, MA) at 2,000 rpm for 30 min with no brake. The layer of cells above the Ficoll containing the lymphocytes was then carefully collected with a sterile transfer pipette and washed twice with 4 ml PBS with 1% BSA. The sample was centrifuged at 1,500 rpm for 7 min. The cell pellet was resuspended in 1 ml PBS with 1% BSA and counted using a Coulter Multisizer Z1 (Beckman Coulter, Fullerton, CA). Multiple washes of the cell pellet were required in order to obtain adequate lymphocyte isolation.

Antibodies

Antibodies (mAb) were purchased from Southern Biotechnology Associates (Birmingham, AL). The antibodies used were the following: Mouse anti-chicken CD8+, Mouse anti-chicken CD4+, Mouse anti-chicken CD3+, and Mouse anti-chicken CD45+. Fluorochromes used for antibody labeling for flow cytometry detection were fluorescein isothiocyanate (FITC) and R-phycoerythrin (R-PE). Both were fluorescent microspheres conjugated to antibodies to allow for specific antibody detection.

Cell Labeling

Working dilutions and conditions for the above mAB conjugates were provided by Southern Biotechnology Associates. Two µl of FITC-conjugated

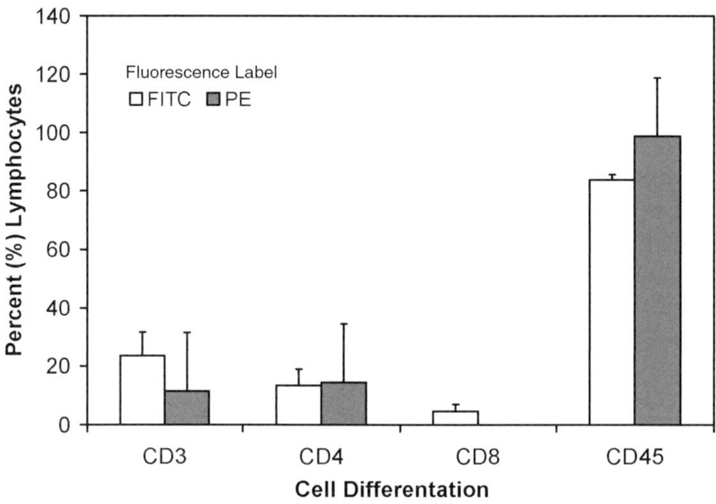

Figure 7.1. Percentage of lymphocytes for each cell differentiation marker and fluorescent label.

mABs was diluted into 20 µl of PBS. For R-PE conjugated mAbs, 0.5 µl was diluted in 20 µl total PBS. These working solutions were then added to approximately 1,000,000 cells. The cells were incubated for 20 min at room temperature in the dark and washed one time with 1X PBS. The cells were centrifuged at 1,500 rpm for 7 min. The pellet was resuspended in 100 µl CAL-LYSE (Caltag Laboratories, Burlingame, CA) and incubated for 10 min at room temperature. Next, 2 ml of ddH20 was added and allowed to incubate for 10 min. The resulting mixture was centrifuged at 1,500 rpm for 7 min and washed once with PBS. The final pellet was resuspended in 250 µl PBS. Combinations used for flow cytometry were: CD4/CD8, CD3/CD8, CD3/CD45, CD4/CD45, and CD8/CD45.

Flow Cytometry and Data Analysis

Flow cytometry analysis was performed on a FACSCalibur flow cytometer (Becton Dickinson, San Jose, CA) with a 488-nm argon laser. Green fluorescence (from FITC) was detected on the FL1 channel (530/30 nm band-pass filter), and orange fluorescence (from R-PE) was detected on the FL2 channel (585/42 nm band-pass filter). Cells were analyzed at up to 20,000 events. Flow cytometry was repeated for the same sample and compared for repeatability.

The Statistical Analysis System (SAS Institute Inc., Cary, NC) was used for all statistical analyses, and assumptions for parametric statistics were examined. Flow cytometry results were compared among antibodies, individual birds, and repeated analysis with nested analysis of variance (ANOVA) models. We considered the among-individual variance components to represent repeatability. Mean percentage for labeled cell types were compared with Duncan's Multiple Range Test. Data not normally distributed or having unequal variances were compared with Kruskal–Wallis nonparametric tests.

RESULTS AND DISCUSSION

Immunophenotyping of CD3+, CD4+, CD8+, and CD45+ was successful for chicken peripheral circulating blood. A total of 85 chickens were used in this experiment over a 6-mo period. Seven to ten individuals were bled and analyzed by flow cytometry per collection day. Gating boxes were created by visually placing a box that encompassed the two separate populations of R-PE or FITC labeled cells as they were appearing on the screen. If needed, these boxes were moved slightly with each run to optimize capturing all the cells. CD8+ cells were labeled only with FITC. No differences were found in the heteroscedasticity between the labels (FITC or R-PE) for each CD marker, except for CD45+ (Levene's test), which was then compared using the nonparametric Kruskal–Wallis test. The percentage of lymphocytes did not differ for each cell differentiation and fluorescent label used for CD4+ that includes a repeated test ($F_{1,100} = 2.10$, $P = 0.15$) (Fig. 7.1). However, fluorescent labels used were different for CD3+ ($F_{1,69} = 36.46$,

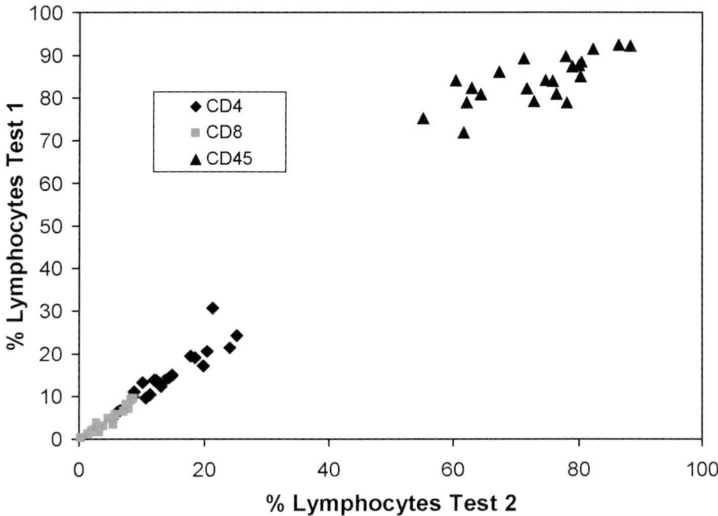

Figure 7.2. Percentage of lymphocytes CD4+, CD8+, and CD45+ compared among individual chickens with repeated flow cytometry measurements, test 1 and test 2. The CD4+ and CD8+ markers used were FITC-labeled and the CD45+ marker was labeled with PE.

$P < 0.0001$) and CD45+ markers ($\chi^2 = 41.3$, $P < 0.001$). Since CD45+ cell markers are essentially on the surfaces of all lymphocytes and, therefore, close to 100%, the PE label appears to be lower than expected, with a mean of 83.8% versus 98.8% for the FITC label. The PE label for CD3+ was also significantly lower than the FITC labeled cell markers, with 11.5% versus 23.6% for the FITC labeled cells. For CD4+ and CD8+, no differences were found between lymphocyte repeated runs, meaning measurements were highly repeatable ($F_{1,100} = 2.47$, $P = 0.15$ and $F_{1,94} = 0.23$, $P = 0.79$, respectively; Fig. 7.2). Antibody markers made from chickens did not cross-react with lymphocytes isolated from Black-billed Magpies and Common Ravens, species not related to chickens. Antibody markers partially reacted with Rock Pigeon cells, and results are inconclusive at this point.

Our work demonstrates the feasibility of using flow cytometry for measurements of peripheral blood in chickens, using anti-chicken antibodies for lymphocyte subpopulations but not in other species of birds unrelated to chickens. Bohls et al. (2006) evaluated peripheral blood mononuclear cells in birds by flow cytometry and found that their protocol could identify CD4+ and CD8+ T lymphocytes in another bird species, the endangered Attwater's Prairie-Chicken (*Tympanuchus cupido*). Although the authors found that a number of monoclonal antibodies specific for domestic chickens did not react with the prairie chicken leukocytes, several were functional and could be reliably used for evaluations that provided dot plots similar to those observed with the analyses of domestic chicken.

Stewart and Nicholson (2000) point out that because each company that produces antibodies may produce them to a different epitope, they may behave differently in function studies. Antibodies can be made to epitopes on the unique portions of restricted proteins, or they can be made to the portion coded for by exons that are common to all family members. In our study, we found differences in the immunophenotyping of lymphocytes using antibodies from two different companies. Based on our results, we recommend that functional studies should use antibodies obtained from the same company.

Comparative immunology of species that vary in susceptibility to WNV will clearly require immunophenotyping techniques that are equally effective for each species compared. Microscopy is still the most common method for measuring lymphocyte differentiation in birds. While we tested whether overlapping epitopes existed for species not related to chickens and found that corvids did not overlap in antibody reaction, future testing for more closely related species such as quail and other gallinaceous species will be required. Susceptibility to WNV may vary greatly in species that appear otherwise more closely related to chickens, such as the Greater Sage-Grouse (*Centrocercus urophasianus*; Naugle et al. 2004, Walker et al. 2004, Clark et al. 2006), which has been found to suffer near 100% mortality to WNV. If sage grouse have an epitope similar to the chicken for the same lymphocyte subpopulations, then this would be the first logical comparison with chickens to a species that varies in susceptibility to WNV. Otherwise, the next step in the technology advancement of flow cytometry for other species will be to develop CD markers in mice or rabbits for species such as Rock Pigeons or Common Ravens. The genetic diversity observed among avian species, especially immunologically, is relatively high compared to mammalian species. Results obtained from lymphocyte population studies provide further evidence to suggest these differences could have a large influence on the variable susceptibility and pathogenesis observed between bird species for a given disease.

Regarding the measurement techniques used, the Coulter Counter® works differently than flow cytometry by detecting changes in electrical conductance of a small aperture as fluid containing cells is drawn through. Coulter counting of the absolute numbers of blood cells does not effectively work for avian cells due to the overwhelming presence of nucleated erythrocytes. In mammals, non-nucleated red blood cells are lysed prior to Coulter counting. However, the development and the possibility of a digital platform for counting large numbers of cells may help in the potential application of Coulter-type counting of avian cells. In humans, immunophenotyping by flow cytometry typically uses estimation of total lymphocytes (e.g., by Coulter counting) in conjunction with measurement of lymphocyte subpopulations by flow cytometry in order to calculate absolute numbers. Our analysis provided percentages of the four lymphocyte CD subpopulations, relative to each other as a percentage of lymphocytes measured by light scatter in the flow cytometer.

Our study demonstrated the importance of adequately isolating lymphocytes for further estimation by flow cytometry of the selected T-cell subpopulations that may be important for fighting WNV infection in birds. We hope our work will encourage other avian pathologists and immunologists to employ flow cytometry in rigorous and controlled experiments for a better understanding of WNV infection in birds. Possible future research will first include experimental infections of chickens and then the further development of flow cytometry techniques for other species. It will be through understanding the mechanisms underlying WNV infection that we will help to elucidate the continuation of WNV in our environment and the impacts of WNV on wild bird populations. The application of immunophenotyping technologies would undoubtedly also impact other aspects of avian medicine, toxicology, disease pathology, and conservation medicine for birds around the world.

ACKNOWLEDGMENTS

We thank the following people for excellence in field and laboratory assistance: C. Hathcock, D. Keller, B. Pearson, L. Marsh, L. Bare, M. Nichols, and A. Thomas. We thank H. Hinjosa for comments on an earlier draft. This research was funded by the Laboratory Directed Research and Development Program through Los Alamos National Security, LLC, operator of the Los Alamos National Laboratory under Contract No. DE-AC52-06NA25396 with the U.S. Department of Energy. This research was approved by the Institutional Animal Care and Use Committee and the Institutional Biosafety Committee at Los Alamos National Laboratory.

LITERATURE CITED

Allison, A. B., D. G. Mead, S. E. J. Gibbs, D. M. Hoffman, and D. E. Stallknecht. 2004. West Nile virus viremia in wild Rock Pigeons. Emerging Infectious Diseases 10:2252–2255.

Altfeld, M., and E. Rosenberg. 2000. The role of CD4 T helper cells in the cytotoxic T lymphocyte response to HIV-1. Current Opinion in Immunology 12: 375–380.

Bohls, R. L., R. Smith, P. J. Ferro, N. J. Silvy, Z. Li, and E. W. Collisson. 2006. The use of flow cytometry to discriminate avian lymphocytes from contaminating thrombocytes. Developmental & Comparative Immunology 30:843–850.

Brinton, M. A. 2001. Host factors involved in West Nile virus replication. Annals of the New York Academy of Science 951:207–219.

Caffrey, C., S. C. R. Smith, and T. J. Weston. 2005. West Nile virus devastates an American Crow population. Condor 107:128–132.

Chu, J. J., and M. Ng. 2004. Interaction of West Nile Virus with αvβ3 integrin mediates virus entry into cells. Journal of Biological Chemistry 279: 54533–54541.

Clark, L., J. Hall, R. McLean, M. Dunbar, K. Klenk, R. Bowen, and C. A. Smeraski. 2006. Susceptibility of Greater Sage-Grouse to experimental infection with West Nile virus. Journal of Wildlife Diseases 42:14–22.

Czekaj, H., E. Samorek-Salamonowicz, W. Kozdrun, D. Bednarek, and K. Krol. 2005. Analysis of T lymphocyte populations in peripheral blood of chickens infected with retroviruses. Medycyna Weterynaryjna 61:703–705.

Deegan, C. S., J. E. Burns, M. Huguenin, E. Y. Steinhaus, N. A. Panella, S. Beckett, and N. Komar. 2005. Sentinel pigeon surveillance for West Nile virus by using lard-can traps at differing elevations and canopy cover classes. Journal of Medical Entomology 42:1039–1044.

Diamond, M. S., B. Shrestha, A. Marri, D. Mahan, and M. Engle. 2003a. B cells and antibody play critical roles in the immediate defense of disseminated infection by West Nile encephalitis virus. Journal of Virology 77:2578–2586.

Diamond, M. S., E. M. Sitati, L. D. Friend, S. Higgs, B. Shrestha, and M. Engle. 2003b. A critical role for induced IgM in the protection against West Nile virus infection. Journal of Experimental Medicine 198:1853–1862.

Doherty, P. C., and J. P. Christensen. 2000. Accessing complexity: the dynamics of virus-specific T cell responses. Annual Review of Immunology 18:561–592.

Erf, G. F., W. G. Bottje, and T. K. Bersi. 1998. CD4, CD8, and TCR defined T-cell subsets in thymus and spleen of 2- and 7-week-old commercial broiler chickens. Veterinary Immunology and Immunopathology 62:339–348.

Furusawa, S., K. Yoshida, K. Kushima, A. Shigeta, H. Horiuchi, and H. Matsuda. 2000. CD3(+)CD4(−)CD8(−) cells present in chicken peripheral blood are regulated by aging and stress. Pp. 315–321 in K. A. Schat (editor), 6th Avian-Immunology-Research-Group Meeting. Cornell University, Ithaca, NY.

Gaehtgens, P., F. Schmidt, and G. Will. 1981. Comparative rheology of nucleated and non-nucleated red blood cells. Pflügers Archiv European Journal of Physiology 390:278–282.

Gibbs, S. E. J., D. M. Hoffman, L. M. Stark, N. L. Marlenee, B. J. Blitvich, B. J. Beaty, and D. E. Stallknecht. 2005. Persistence of antibodies to West Nile virus in naturally infected Rock Pigeons (*Columba livia*). Clinical and Diagnostic Laboratory Immunology 12:665–667.

Halevy, M., Y. Akov, D. Bennathan, D. Kobiler, B. Lachmi, and S. Lustig. 1994. Loss of active neuroinvasiveness in attenuated strains of West Nile virus: pathogenicity in immunocompetent and SCID mice. Archives of Virology 137:355–370.

Hayday, A. C. 2000. Gamma delta cells: a right time and a right place for a conserved third way of protection. Annual Review of Immunology 18:975–1026.

Kindt, T. J., R. A. Goldsby, and B. A. Osborne. 2007. Immunology. W. H. Freeman and Company, New York, NY.

Kinney, R. M., C. Y. H. Huang, M. C. Whiteman, R. A. Bowen, S. A. Langevin, B. R. Miller, and A. C. Brault. 2006. Avian virulence and thermostable replication of the North American strain of West Nile virus. Journal of General Virology 87:3611–3622.

Kliger, C. A., A. E. Gehad, R. M. Hulet, W. B. Roush, H. S. Lillehoj, and M. M. Mashaly. 2000. Effects of photoperiod and melatonin on lymphocyte activities in male broiler chickens. Poultry Science 79:18–25.

Komar, N., S. Langevin, S. Hinten, N. Nemeth, E. Edwards, D. Hettler, B. Davis, R. Bowen, and M. Bunning. 2003. Experimental infection of North American birds with the New York 1999 strain of West Nile virus. Emerging Infectious Diseases 9:311–322.

Kulkarni, A. B., A. Muellbacher, and R. V. Blanden. 1991. Functional analysis of macrophages B cell and spenic dendritic cells as antigen-presenting cells in West Nile virus-specific murine T lymphocyte proliferation. Immunology and Cell Biology 69:71–80.

LaDeau, S. L., A. M. Kilpatrick, and P. P. Marra. 2007. West Nile virus emergence and large-scale declines of North American bird populations. Nature 447:710–713.

Lanciotti, R. S., J. T. Roehrig, V. Deubel, J. Smith, M. Parker, K. Steele, B. Crise, K. E. Volpe, M. B. Crabtree, J. H. Scherret, R. A. Hall, J. S. MacKenzie, C. B. Cropp, B. Panigrahy, E. Ostlund, B. Schmitt, M. Malkinson, C. Banet, J. Weissman, N. Komar, H. M. Savage, W. Stone, T. McNamara, and D. J. Gubler. 1999. Origin of the West Nile virus responsible for an outbreak of encephalitis in the northeastern United States. Science 286:2333–2337.

Langevin, S. A., M. Bunning, B. Davis, and N. Komar. 2001. Experimental infection of chickens as candidate sentinels for West Nile virus. Emerging Infectious Diseases 7:726–729.

Lillehoj, H. S. 1991. Lymphocytes involved in cell-mediated immune responses and methods to assess cell-mediated immunity. Poultry Science 70:1154–1164.

Liu, Y., R. V. Blanden, and A. Mullbacher. 1989. Identification of cytolytic lymphocytes in West Nile virus-infected murine central nervous system. Journal of General Virology 70:565–574.

McLean, R. G. 2006. West Nile virus in North American birds. Ornithological Monographs. American Ornithologists Union, Washington, DC.

Naugle, D. E., C. L. Aldridge, B. L. Walker, T. E. Cornish, B. J. Moynahan, M. J. Holloran, K. Brown, G. D. Johnson,

E. T. Schmidtmann, R. T. Mayer, C. Y. Kato, M. R. Matchett, T. J. Christiansen, W. E. Cook, T. Creekmore, R. D. Falise, E. T. Rinkes, and M. S. Boyce. 2004. West Nile virus: pending crisis for Greater Sage-Grouse. Ecology Letters 7:704–713.

Nemeth, N. M., and R. A. Bowen. 2007. Dynamics of passive immunity to West Nile virus in domestic chickens (*Gallus gallus domesticus*). American Journal of Tropical Medicine and Hygiene 76:310–317.

Norris, K., and M. R. Evans. 2000. Ecological immunology: life history trade-offs and immune defense in birds. Behavioral Ecology 11:19–26.

Paramithiotis, E., L. Tkalec, and M. J. H. Ratcliffe. 1991. High levels of CD45 are coordinately expressed with CD4 and CD8 on avian thymocytes. Journal of Immunology 147:3710–3717.

Picker, L., and V. Maino. 2000. The CD4 T cell response to HIV-1. Current Opinion in Immunology 12:381–386.

Potter, P. 2004. North American birds and West Nile virus. Emerging Infectious Diseases 10:1518–1519.

Samuel, M. A., and M. S. Diamond. 2006. Pathogenesis of West Nile virus infection: a balance between virulence, innate and adaptive immunity, and viral evasion. Journal of Virology 80:9349–9360.

Shrestha, B., and M. S. Diamond. 2004. Role of CD8+ T cells in control of West Nile virus infection. Journal of Virology 78:8312–8321.

Shrestha, B., M. A. Samuel, and M. S. Diamond. 2006. CD8(+) T cells require perforin to clear West Nile virus from infected neurons. Journal of Virology 80:119–129.

Sitati, E. M., and M. S. Diamond. 2006. CD4(+) T-cell responses are required for clearance of West Nile virus from the central nervous system. Journal of Virology 80:12060–12069.

Stewart, C. C., and J. K. A. Nicholson (editors). 2000. Immunophenotyping. Wiley-Liss, New York, NY.

Suzuki, Y., T. Ito, T. Suzuki, R. E. Holland Jr., T. M. Chambers, M. Kiso, H. Ishida, and Y. Kawaoka. 2000. Sialic acid species as a determinant of the host range of influenza A viruses. Journal of Virology 74:11825–11831.

Walker, B. L., D. E. Naugle, K. E. Doherty, and T. E. Cornish. 2004. Outbreak of West Nile virus in Greater Sage-Grouse and guidelines for monitoring, handling, and submitting dead birds. Wildlife Society Bulletin 32:1000–1006.

Wang, T., E. Scully, Z. N. Yin, J. H. Kim, S. Wang, J. Yan, M. Mamula, J. F. Anderson, J. Craft, and E. Fikrig. 2003a. IFN-gamma-producing gamma delta T cells help control murine West Nile virus infection. Journal of Immunology 171:2524–2531.

Wang, Y., M. Lobigs, E. Lee, and A. Mullbacher. 2003b. CD8+ T cells mediate recovery and immunopathology in West Nile virus encephalitis. Journal of Virology 77:13323–13334.

Ward, M. R., D. E. Stallknecht, J. Willis, M. J. Conroy, and W. R. Davidson. 2006. Wild bird mortality and West Nile virus surveillance: biases associated with detection, reporting, and carcass persistence. Journal of Wildlife Diseases 42:92–106.

Yaremych, S. A., R. E. Warner, P. C. Mankin, J. D. Brawn, A. Raim, and R. Novak. 2004. West Nile virus and high death rate in American Crows. Emerging Infectious Diseases 10:709–711.

CHAPTER EIGHT

Zoonotic Diseases

WHAT ORNITHOLOGISTS AND BIRD BANDERS SHOULD KNOW

Ornithological Council

Highly pathogenic avian influenza H5N1 ("HPAI H5N1") first made news in 2004 and seemed to dominate headlines for several years. The alarmism belies the fact that the impact to human health has been slight. Though human outbreaks have been occurring since 1997 (WHO 2005), only 500 human cases, including 294 deaths, were reported to the World Health Organization from 2003 through July 2010 (WHO 2010). Though there have been several confirmed cases of human-to-human transmission resulting from close, prolonged contact between family members or from an infected individual to a health care worker, nearly all other human cases—which have occurred primarily in healthy adults and children—are attributed to direct handling of infected poultry, consumption of undercooked poultry products, or contact with virus-contaminated surfaces or materials used in handling poultry (Writing Committee 2006). To date, only seven human cases of HPAI H5N1 infection appear to be related to contact with wild birds, and these resulted from the plucking of feathers from dead swans in Azerbaijan. It is not clear that all seven cases resulted from contact with the dead birds, or if one or more cases resulted from contact with those who handled the dead birds (WHO 2006, Tsiodras et al. 2008).

At least in the United States, HPAI H5N1 has faded from the news, but it was not the first avian wildlife disease to cause substantial concern and it will not be the last. Only a few years earlier, West Nile virus (WNV) commanded the public's attention when it first appeared in the United States. First isolated in 1937 in Uganda, WNV has caused outbreaks in Israel (1951–1954), France (1962, 2000), and South Africa (1974). In 1999 it reached the United States, where researchers—and their universities, government research agencies, and other research organizations—became concerned about the risk to field biologists, students, and others. Perhaps out of an abundance of caution and spurred by constant media attention, one university canceled field research and field biology classes that involved handling birds. The following was soon determined:

1. Most people who are infected with WNV do not develop any type of illness.

2. It is estimated that 20% of the people who become infected will develop West Nile fever and will experience mild symptoms, including fever, headache, and body aches, occasionally with a skin rash on the trunk of the body and swollen lymph glands.

Ornithological Council. 2012. Zoonotic diseases: what ornithologists and bird banders should know. Pp. 91–100 *in* E. Paul (editor). Emerging avian disease. Studies in Avian Biology (vol. 42), University of California Press, Berkeley, CA.

3. About one of every 150 infected persons becomes seriously ill with central nervous system infection (encephalitis and/or meningitis; CDC 2010).

4. For young/healthy researchers who are not immunocompromised, West Nile virus is unlikely to cause much more than a mild illness—typically "flu-like symptoms."

The Ornithological Council—a consortium of 12 scientific ornithological societies in the Western Hemisphere—consulted with a number of experts to compile a fact sheet about the risks of HPAI H5N1 and WNV for ornithologists and bird banders and to provide the most up-to-date occupational safety and animal welfare recommendations for those handling live birds, carcasses, or tissues that are potentially infected with WNV or HPAI H5N1.

Ornithologists handling wild birds may also be exposed to other zoonotic pathogens including *Salmonella* spp. and *Chlamydia psittaci* (also known as ornithosis or psittacosis). Because ornithologists and bird banders handle live birds, prepare specimens, and handle blood and other tissues of avian origin, they need to understand the means of transmission of zoonotic pathogens and know effective means to protect themselves and the birds they study.

The measures that should be taken to avoid contracting a zoonotic disease and to avoid transmitting it to others should be commensurate with the extent of the risk and of the consequences of contracting the disease. Preventive measures can be burdensome and interfere with research techniques, especially under field conditions. However, if encountering a pathogen that has the potential to cause serious disease, more extensive measures are warranted even if burdensome, uncomfortable, or costly.

Check frequently for updates of this fact sheet as new zoonotic diseases emerge or as conditions or degree of risk may change. Updates will be posted on BIRDNET, the website of the Ornithological Council (www.nmnh.si.edu/birdnet).

AVIAN INFLUENZA

The Basics

Various avian influenza viruses are found in wild birds in virtually every country, including the United States. The subtypes are named for the 16 hemagglutinin (H) and 9 neuraminidase (N) proteins on the viral surface. The avian influenza virus of recent concern is designated as Highly Pathogenic Avian Influenza (HPAI) subtype H5N1 genotype Z, which first appeared in Asia in 2002. Other avian influenza viruses are designated "LPAI" for low pathogenicity. The degree of pathogenicity is established through testing methods developed by the World Health Organization and the International Office of Epizootics (www.oie.int). The pathogenicity designation pertains only to the behavior of the virus in domestic poultry; a virus may not behave the same way in wild birds.

Many avian influenza viruses normally circulate as gastrointestinal infections in wild birds, causing little or no illness or mortality (Webster et al. 1992). The H5N1 strain of HPAI has affected 152 species in 14 orders of wild birds and has caused mortality in 115 of those species (USGS 2010). Bird species in many families appear to be susceptible to infection, but because cool, wet conditions favor the persistence of the virus, and because the virus is shed in feces that contaminate their aquatic habitats, it appears that waterbirds, especially ducks and geese, are the most commonly infected wild birds (Causey and Edwards 2008).

Studies have been conducted to determine if wild birds can be healthy carriers of HPAI H5N1 virus, to study the role of healthy carriers in the spread of the disease, and to gather information on the routes and periods of migration of the infected wild birds. It has proved difficult to find healthy, infected birds. In 2006, none of the 39,143 wild birds of 150 species sampled in Europe were found to be infected (Pittman et al. 2007). In a study that sampled 13,000+ live migratory birds in China, HPAI H5N1 was detected only six times (Chen et al. 2006). Of 862 live birds tested across the western Mongolian flyway, including 430 live birds (of 55 species) found on Erhel Lake in Mongolia, where a mass mortality event killed 100 birds, none tested positive for the virus (WCS 2005).

Where the Highly Pathogenic Form of H5N1 Avian Influenza Is Found

The World Animal Health Information Database, compiled by the World Organization for Animal

Health (OIE), provides the official disease status for each country (www.oie.int/wahis/public.php?page=home). The website of the USDA Animal and Plant Health Inspection Service, National Center for Import and Export (www.aphis.usda.gov/import_export/animals/animal_disease_status.shtml) lists countries that the agency has recognized as free of certain diseases. Early reports, often prior to official confirmation, can be found on ProMed (www.promedmail.org).

Precautions to Take When Working in the Field

Infected birds shed flu virus in their saliva, nasal and tracheal secretions, and feces. The virus has also been detected on the feathers of wild birds (Delogu et al. 2010). Although "high pathogenicity" is generally associated with the rapid onset of severe illness and high mortality, it has been confirmed in laboratory tests and through sampling of wild birds that some infected birds can be appear healthy (Kou et al. 2005, Chen et al. 2006).

It is not yet known if some species are more likely to be healthy carriers or are more efficient at transmitting the virus than are other species. Most dead, wild birds found to have been infected with H5N1 are waterfowl species, but this may reflect the fact that the carcasses of large birds are more readily noticed than are the carcasses of small bird species, which will likely decompose or be scavenged before they are found. Surveillance of live wild birds has focused on waterfowl; in the European Union, 62% of the birds tested were waterfowl. Despite the acknowledged research bias and the lack of published data, however, experts think it likely that Anseriformes are more susceptible to HPAI H5N1 infection than are other taxa, although in experimentally inoculated birds, mortality was higher in gallinaceous birds, finches, geese, Emus (*Dromaius novaehollandiae*), and Budgerigars (*Melopsittacus undulatus*) (Perkins and Swayne 2003). Cool, wet conditions favor persistence of the virus (Causey and Edwards 2008), and the virus may be concentrated in the habitat used by species that congregate in large numbers, as do waterfowl.

When working in countries or regions where H5N1 has been confirmed, or along pathways used by birds migrating to, from, or through regions where H5N1 has been confirmed, assume that the birds you handle may have been or may be shedding virus. Whatever the risk of encountering the virus and contracting the disease, the disease is difficult to treat and the mortality rate is fairly high. Therefore, to protect yourself, do the following:

1. Avoid unprotected contact with feces, secretions, blood, and fluids. Wear protective clothing including shoe covers or rubber boots, eye protection, and gloves. If you cannot do so, decontaminate and clean yourself immediately after exposure, using a detergent-based cleanser. Disinfect or dispose of protective clothing after use.

2. Learn to remove gloves and protective clothing in a manner that avoids skin contact; consult your safety officer or safety manual.

3. Wash hands immediately with soap and water. Use a respirator or mask to avoid inhalation of aerosolized droplets; otherwise, work upwind of birds to avoid inhaling aerosolized fecal material, feathers, and dander. After handling birds, use detergent-based cleansers to wash hands, equipment, and clothing. Alcohol (70%) or alcohol-based cleansers or diluted household bleach (10% strength) will also kill the virus.

4. Avoid eating or drinking while handling birds or bird parts.

5. Consider having antiviral medications on hand. Ask your physician if you should take these medications on a prophylactic basis before you begin working in a country or region where H5N1 has been confirmed or along pathways used by birds migrating to, from, or through countries or regions where H5N1 occurs. Any influenza strain can become resistant to one or more drugs; genetically distinct H5N1 subtypes have already been found in Asia and some antivirals may be more effective for some subtypes than for others. Be sure to check current health information from a credible source, such as the Centers for Disease Control, for both country disease status and antiviral recommendations, and seek a prescription for the appropriate medication from your physician.

6. Consider vaccines, if they are available. The National Institutes of Health began testing a vaccine in clinical trials in April 2005. The

current CDC recommendations to travelers to and residents of HPAI H5N1 countries do not include vaccination, but it is recommended to avoid contact with domestic and wild birds. As ornithologists and banders will, of course, handle wild birds, a consultation with your physician or infectious disease specialist about the use of an appropriate vaccine is recommended.

The university or research institution may attempt to restrict field research. Know the disease status of the countries where you intend to work and be prepared to inform the risk management office (or, in the United States, the Institutional Animal Care and Use Committee, which, in many universities, performs risk management functions) along with the precautions you plan to take. It is the researcher's responsibility to know the recommended precautions and to make arrangements to obtain and use the appropriate materials, such as disinfectants, gloves, and eye protection.

Ornithologists should know that the USDA restricts imports of birds and bird products (defined by the USDA as "anything that was once a bird or a part of a bird") from countries where *any* HPAI avian influenza is known to exist. Permits for such imports are conditioned upon the importer promising, in the permit application, to treat the specimens and tissues with a USDA-approved method to inactivate the virus, and importers must supply a certification, upon arrival in the United States, that the imported materials have, in fact, been treated in accordance with the methods delineated in the permit. The Ornithological Council has published detailed guides to the import of birds and bird products (www.nmnh.si.edu/birdnet/permits.html). If your institution has a permit that does not include imports from the countries where *any* HPAI is found, or if you do not provide a copy of the permit and the certification of treatment upon arrival in the United States, the specimens or samples will be refused entry and will probably be confiscated and destroyed.

Some universities and museums recommend or require a period of quarantine for biologists returning from fieldwork in countries or regions where the current HPAI H5N1 strain is present. In addition to the fact that very few cases of human-to-human transmission have been confirmed, and these few cases have occurred only after close, prolonged contact, recent research suggests that this precaution may be scientifically unwarranted. Research shows that person-to-person transmission is unlikely because the virus preferentially attaches to cell types that are found in the lower respiratory tract. If the virus cannot replicate in the upper respiratory tract, it is difficult to transmit through coughing and sneezing, which is the most common means of viral transmission among humans (Shinya et al. 2006, van Riel et al. 2006). In late 2011, it was reported in the press that a research group in The Netherlands had succeeded in creating a genetically modified H5N1 virus that could be transmitted through the air, facilitating much easier transmission, including transmission between infected humans (Grady and McNeil 2011). However, this variant, which entailed multiple mutations, exists only in the laboratory. There has been concern that the virus will mutate or will reassort with other viruses that circulate among humans, and will acquire characteristics that make it easier to transmit between humans. However, deliberate manipulation of the H5N1 genome that produced mutations and reassortments with other avian influenza viruses that humans contract failed to produce characteristics that increased transmissibility (Maines et al. 2006). Before you return, any clothes worn in the field should be laundered with detergent and should not be worn again until they have been laundered. Field equipment should be disinfected after use, as described above.

Precautions to Take When Working
in the Laboratory

Ornithologists preparing specimens or working with blood or tissue from fresh birds should be aware that the virus will remain viable in dead birds for several days, particularly in cool or wet climates. Freezing does not kill viruses; those working with thawed tissue from birds originating in countries or regions where HPAI occurs should take appropriate precautions. The USDA-approved treatment methods (as described in the OC Permit Guide to Import of Bird Specimens) will inactivate the virus. If you have imported birds from HPAI countries, you will have been required to use one of these methods prior to import and will have inactivated the virus. Nonetheless, it is recommended by the World Health Organization that work be conducted in a laboratory that meets Biosafety Level 2

(BSL2) conditions. These standards are found in the manual for Biosafety in Microbiological and Biomedical Laboratories (BMBL).

A university or research institution's risk management office may attempt to impose restrictions on work involving materials imported from countries where HPAI H5N1 occurs because the BMBL states that BSL3 is appropriate when "work is done with indigenous or exotic agents with a potential for respiratory transmission, and which may cause serious and potentially lethal infection." Should this occur, the researcher can explain to the university that the import permit required pre-import inactivation of the virus using a USDA-approved method, so the materials do not contain infectious agents. In the event that H5N1 is confirmed in wild or domestic birds in the United States, these pre-import treatment methods would, of course, not be required. It would then be possible to bring tissue containing H5N1 virus into the laboratory. Only then would BSL3 conditions be required, and then only if you choose not to treat the material so as to inactivate the virus and if the manner of manipulation of the tissue would be likely to result in aerosolization.

WEST NILE VIRUS

The Basics

West Nile is an insect-borne flavivirus commonly found in Africa, western Asia, and the Middle East, and, since 1999, in the Western Hemisphere. In North America, it has been detected in at least 48 species of mosquitoes and over 250 species of birds (USGS 2010). It is now found in every state except Alaska and Hawaii.

Precautions to Take When Working in the Field

Although there are no documented cases of ornithologists or bird banders contracting WNV from handling living or dead birds, it is also the case that there has been no surveillance of ornithologists or bird banders to determine the presence/absence or prevalence of the disease. Even if such surveillance were to be implemented, it would be difficult to know if the disease had been contracted through contact with bird feces or saliva or if it had been contracted from an insect bite—at the research site or elsewhere.

It has been confirmed that WNV may be shed from the cloacal and oral cavities (Komar et al. 2002). Therefore, contact with droppings, dropping-contaminated feathers, or the cloaca may result in exposure to WNV.

1. Be sure to have antiseptic (not antibacterial or antimicrobial) wipes or gels available for hand washing and first aid for cuts or punctures sustained while handling birds. Using wipes after handling each bird will protect both the researcher and the birds subsequently handled by the researcher.

2. Avoid contact with bird feces.

3. If bitten by a bird, wash hands (when possible) or use antiseptic (not antibacterial or antimicrobial) wipes or even a small amount of fresh bleach.

4. Extra care should be taken to avoid needle sticks when taking blood samples. Public health officials consider gloves to be an appropriate precaution but ornithologists rarely wear gloves when handling birds, particularly in the field. If gloves are worn, they should be changed or decontaminated with 70% ethanol or other appropriate substance after handling each bird to avoid transmission from one bird to another. Be familiar with proper glove removal, which entails avoiding contact with the skin, and disposal. Other barrier protections such as goggles and masks are standard precautions against inadvertent exposure to droplets and fluids. However, goggles and masks are probably disproportionate to the nature and extent of the risk posed by this particular pathogen.

5. Take same reasonable precautions to minimize risks—of various diseases—posed by mosquito bites. Reasonable measures include protective clothing (long sleeves, long pants, socks), and the use of DEET or other insect repellants, with repeated applications over time. For detailed information about the proper use of DEET and a summary of a recent assessment of the efficacy and safety of DEET, see the appendix.

Precautions to Take When Working in the Lab

As of February 2003, there have been only two documented cases of researchers contracting West Nile Virus in the course of conducting research. Both cases involved microbiologists. One was infected from an accidental needle puncture in the finger while working with live virus, and the other was infected through an accidental scalpel cut while performing a necropsy on a dead Blue Jay (*Cyanocitta cristata*; CDC 2002).

It is best to assume that any specimen or tissue sample that has not been treated with a method known to kill the virus could be infectious and to take proper precautions at all times. Neither refrigeration nor freezing will kill the virus. Assume that thawed tissue or specimens from birds could contain live virus. The virus can remain viable in dead birds for several days. Do the following:

1. Take care to avoid scalpel cuts and punctures. If they occur, cleanse the area promptly and thoroughly, apply antiseptic and report the incident to a supervisor. If signs of illness occur within two weeks of exposure, seek prompt medical evaluation and consult with public health authorities.

2. Take standard measures to minimize exposure to fluids or tissues during handling of potentially infected tissue including barrier protections such as gloves, masks, and eyewear; properly use and dispose of needles, scalpels, and other sharp instruments; and minimize the generation of aerosols (such as vigorous spraying of water on carcasses or work surfaces). At least when preparing specimens, ornithologists rarely wear gloves. Given the nature of the risk, an appropriate alternative is taking care to avoid scalpel cuts and punctures, and washing promptly if a cut or puncture wound occurs, followed by the use of an antiseptic or 70% alcohol. However, gloves should be worn if the skin is broken. Aerosolization rarely occurs when preparing specimens, but procedures that could possibly produce aerosols or splashing can be conducted under a biosafety hood.

3. Avoid touching anything but the materials involved in the procedure. Touching equipment, phones, wastebaskets, or other surfaces may cause contamination. Decontaminate any surfaces that were touched. Be aware of touching the face, hair, and clothing as well.

OTHER ZOONOTIC PATHOGENS

Wild birds may carry other diseases to which ornithologists and banders are susceptible, and an ornithologist or a bander may easily transfer some avian pathogens from one bird to another. According to the USGS Field Manual of Wildlife Disease, "As a group, bacterial diseases pose greater human health risks than viral diseases of wild birds. Of the diseases addressed in this section, chlamydiosis, or ornithosis, poses the greatest risk to humans. Avian tuberculosis can be a significant risk for humans who are immunocompromised. Salmonellosis is a common, but seldom fatal, human infection that can be acquired from infected wild birds." However, other avian diseases rarely cause illness, much less serious illness in humans, and rarely, if ever, result in death.

According to the CDC, chlamydiosis (also known as ornithosis or psittacosis) is characterized by fever, chills, headache, myalgia, and a dry cough, with pneumonia often evident on chest X-ray. Severe pneumonia requiring intensive-care support, endocarditis, hepatitis, and neurologic complications occasionally occur. Most people recover from salmonellosis in a week or less without medication, though the severe dehydration that can occur can be dangerous and may require hospitalization. Human fatalities from bacterial diseases are rare due to the availability of antibiotics. However, there have been several severe cases among wildlife biologists (Wobeser and Brand 1982).

The level of precaution should be commensurate with the level of risk to the individual handling the bird and to other birds. In most situations, then, hand washing and disinfecting of equipment and holding devices should be adequate.

It is always helpful to recognize the signs of illness in a bird, but because birds can harbor pathogens without showing overt signs of illness, do not assume that the absence of signs indicates the absence of a pathogen. A researcher who becomes ill after handling wild birds should inform the physician of the possible exposure to a zoonotic pathogen.

PRECAUTIONS AGAINST TRANSMISSION OF ALL PATHOGENS TO BIRDS AND OTHER WILDLIFE

To prevent transmission *of any pathogen* as a result of handling by researchers:

1. Do not reuse contaminated bags, boxes, or other holding/carrying devices or other devices used to restrain birds during

processing. The North American Banding Council manual states, "Launder bird bags frequently, as they must be kept clean," and "If a diseased bird is caught, it is extremely important to put that bag aside until it has been washed and disinfected." However, as it is not possible to determine if a bird is shedding virus, the better practice would be to carry an ample supply of bags or other holding/carrying devices so that no bag or other holding device is used more than once before laundering. Viruses can survive at cool temperatures for days, weeks, or even longer. Wash bags with hot water, detergent, and/or household bleach before reuse.

2. When preparing specimens in the field, place waste material in a biosafety bag, seal it, and burn it, or carry it out with you and burn it later. Never reuse needles, scalpel blades, calipers, rulers, banding pliers, or other equipment that touches any part of a bird unless the equipment is decontaminated with a freshly prepared 10% bleach or 70% alcohol solution or alcohol wipes after use on each individual. The National Veterinary Standards Laboratory of the U.S. Department of Agriculture, which approves pre-import treatment methods for materials of avian origin, confirmed that 70% alcohol will kill viruses.

3. Disinfect your hands after handling each bird. Disinfectant hand wipes can be used if washing with soap and water is not possible.

4. For field surgeries, aseptic technique is discussed at length in *Guidelines to the Use of Wild Birds in Research* (Fair et al. 2010).

WHAT ORNITHOLOGISTS AND BANDERS CAN DO IN THE EVENT OF EMERGENT AVIAN DISEASE OR DISEASE OUTBREAKS

Ornithologists and banders can and should develop relationships with their state or provincial health and agriculture departments. For a comprehensive list of state agencies in the United States, see www.pandemicflu.gov/state/statecontacts.html. Should emerging infectious avian diseases arrive in your country, state, or province, or should disease outbreaks occur, you will be prepared to help persuade your state officials to continue

monitoring wildlife after occurrence is confirmed, can help to share accurate scientific information about wild birds with these agencies and with the public, and can help address calls from the public or from government officials to cull wild birds. Every international and national agriculture and public health organization, including the World Health Organization and the United Nations Food and Agriculture Organization, has concluded that culling of wild birds or destruction of their habitat—such as the draining of wetlands—is neither practical nor feasible from logistical, environmental, public health, and biodiversity points of view. In fact, the FAO points out that the attempt to cull or the destruction of habitat could result in the dispersion of birds and if those birds were infected, dispersion would result in spread of the virus to a wider area.

Ornithologists can also serve as experts to provide information to the general public and the media, but should be careful to avoid speculating about how or how quickly the disease might spread; if, when, and how it might arrive in the Western Hemisphere or about any other matter about which information is lacking or incomplete. Speculation can lead to calls for inappropriate measures.

Ornithologists, banders, and bird observatories can greatly extend biosurveillance capacity. Contact information for U.S. organizations already involved in biosurveillance is listed below:

Global Avian Influenza Network for Surveillance
www.gains.org

Institute for Bird Populations
www.birdpop.org

Landbird Migration Monitoring Network
of North America
www.klamathbird.org/lammna

Smithsonian Migratory Bird Center
www.si.edu/smbc

USGS National Wildlife Health Center
www.nwhc.usgs.gov

Submitting Samples for Analysis

Unpublished research presented by Ron A. M. Fouchier at the Food and Agriculture Organization's International Scientific Conference on Avian Influenza and Wild Birds (Rome, 30–31 May 2006) and the Wildlife Disease Association (Storrs, CT, 2006) suggests that cloacal swabs may not be effective for detecting HPAI H5N1. Dabbling ducks experimentally inoculated with H5N1 did not shed the virus in feces and thus cloacal swabs may not be effective in detecting the virus. However, the virus was found in throat swabs in high quantities. If taking cloacal swabs specifically to test for presence of HPAI H5N1, consult the current literature on this subject. The choice of cloacal swabs versus tracheal swabs may also be determined in part by the species and the impact of this procedure on the birds, as well as the level of training and experience of the field staff.

If you collect cloacal or tracheal swabs during the course of your own research, you may want to have the samples analyzed if you have the funding to do so. The U.S. federal agencies are not currently accepting samples for analysis. However, if the specimens are collected according to the protocols in the Interagency Avian Strategic Plan, Attachment 9 (www.usda.gov/documents/wildbirdstrategicplanpdf.pdf) and if the specimens are analyzed by a laboratory with the National Animal Health Laboratory Network (www.nahln.org), you are welcome to submit data to the HPAI Early Detection Data System (HEDDS, wildlifedisease.nbii.gov/ai/index.jsp). Protocols for collecting and shipping avian carcasses for diagnostic evaluation are outlined in Attachment 8 of the same document (www.usda.gov/documents/wildbirdstrategicplanpdf.pdf).

Do not submit samples directly to the USGS National Wildlife Health Center or the USDA National Veterinary Standards Laboratory. Make arrangements in advance to submit samples to your state. A full list of state agriculture, wildlife, and public health departments can be found here: www.pandemicflu.gov/state/statecontacts.html.

In Canada, sampling must be coordinated through Environment Canada, even if the bander or ornithologist is participating in an organized banding program. If you find a dead bird, contact the Canadian Cooperative Wildlife Health Center. Submission forms can be found here: www.ccwhc.ca/wildlife_submission_forms_regions.php.

In Mexico, contact the Avian Flu Commission (www.huitzil.net/nuevo_sitio/PlandetrabajoGripeAviar.pdf).

ADDITIONAL RESOURCES

USGS National Wildlife Health Center has made available its course materials on avian zoonotic disease: www.nwhc.usgs.gov/outreach/avian_zoonotic_course.jsp.

LITERATURE CITED

Causey, D., and S. Edwards. 2008. Ecology of avian influenza virus in birds. Journal of Infectious Disease 197(Suppl.1):S29–S33.

Centers for Disease Control. 2002. CDC morbidity and mortality weekly review (20 December 2002).

Centers for Disease Control. 2010. West Nile virus: questions and answers. <www.cdc.gov/ncidod/dvbid/westnile/qa/symptoms.htm> (1 June 2010).

Chen, H., G. J. Smith, K. S. Li, J. Wang, X. H. Fan, J. M. Rayner, D. Vijaykrishna, J. X. Zhang, L. J. Zhang, C. T. Guo, C. L. Cheung, K. M. Xu, L. Duan, K. Huang, K. Qin, Y. H. Leung, W. L. Wu, H. R. Lu, Y. Chen, N. S. Xia, T. S. Naipospos, K. Y. Yuen, S. S. Hassan, S. Bahri, T. D. Nguyen, R. G. Webster, J. S. Peiris, and Y. Guan. 2006. Establishment of multiple sublineages of H5N1 influenza virus in Asia: implications for pandemic control. Proceedings of the National Academy of Sciences USA 103:2845–2850.

Delogu, M., M. A. De Marco, L. Di Trani, E. Raffini, C. Cotti, S. Puzelli, F. Ostanello, R. G. Webster, A. Cassone, and I. Donatelli. 2010. Can preening contribute to influenza A virus infection in wild waterbirds? PLoS ONE 5:e11315. doi:10.1371/journal.pone.0011315.

Fair, J., E. Paul, and J. Jones. 2010. Guidelines to the use of wild birds in research. Ornithological Council, Washington, DC. Grady, D. and D.G. McNeil, Jr. Debate persists on deadly flu made airborne. New York Times: 27 December 2011; p.A-1.

Komar, N., R. Lanciotti, R. Bowen, S. Langevin, and M. Bunning. 2002. Detection of West Nile virus in oral and cloacal swabs collected from bird carcasses. Emerging Infectious Diseases 8:741–742.

Kou, Z., F. M. Lei, J. Yu, Z. J. Fan, Z. H. Yin, C. X. Jia, K. J. Xiong, Y. H. Sun, X. W. Zhang, X. M. Wu, X. B. Go, and T. X. Li. 2005. New genotype of avian influenza H5N1 viruses isolated from Tree Sparrows in China. Journal of Virology 79:15460–15466.

Maines, T. R., L.-M. Chen, Y. Matsuoka, H. Chen, T. Rowe, J. Ortin, A. Falco, N. T. Hien, L. Q. Mai, E. R. Sedyaningsih, S. Harun, T. M. Tumpey, R. O. Donis, N. J. Cox, K. Subbarao, and J. M. Katz. 2006. Lack of transmission of H5N1 avian-human reassortant influenza viruses in a ferret model. Proceedings of the National Academy of Sciences USA 103:12121–12126.

Perkins, L. E. L., and D. E. Swayne. 2003. Comparative susceptibility of selected avian and mammalian species to a Hong Kong-origin H5N1 high-pathogenicity avian influenza virus. Avian Diseases 47:956–967.

Pittman , M., A. Laddomada, R. Freigofas, V. Piazza, A. Brouw, and I. H. Brown. 2007. Surveillance, prevention, and disease management of avian influenza in the European Union. Journal of Wildlife Diseases 43(Supplement):S64–S70.

Shinya, K., E. Masahito, S. Yamada, M. Ono, N. Kasai, and Y. Kawaoka. 2006. Influenza virus receptors in the human airway. Nature 440:435–436.

Sturm-Ramirez, K. M., D. J. Hulse-Post, E. A. Govorkova, J. Humberd, P. Seiler, P. Puthavathana, C. Buranathai, T. D. Nguyen, A. Chaisingh, H. T. Long, T. S. P. Naipospos, H. Chen, T. M. Ellis, Y. Guan, J. S. M. Peiris, and R. G. Webster. 2005. Are ducks contributing to the endemicity of highly pathogenic H5N1 influenza? Journal of Virology 79:11269–11279.

Tsiodras, S., T. Kelesidis, I. Kelesidis, U. Bauchinger, and M. Falagas. 2008. Human infections associated with wild birds. Journal of Infection 56:83–98.

U.S. Geological Survey. 2010. List of species affected by H5N1. <www.nwhc.usgs.gov/disease_information/ avian_influenza/affected_species_chart.jsp> (30 June 2010).

van Riel, D. V. J., E. de W. Munster, G. F. Rimmelzwaan, R. A. M. Fouchier, A. D. M. E. Osterhaus, and T. Kuiken. 2006. H5N1 virus attachment to lower respiratory tract. Science 312:399.

Webster, R. G., W. J. Bean, O. T. Gorman, T. M. Chambers, and Y. Kawaoka. 1992. Evolution and ecology of influenza A viruses. Microbiological Reviews 56:152–179.

WCS. 2005. Joint Wildlife Conservation–Mongolia Avian Influenza Survey. Final Report to the United Nations Food and Agricultural Organization. Wildlife Conservation Society, Washington, DC.

Wobeser, G., and C. J. Brand. 1982. Chlamydiosis in two biologists investigating disease occurrences in wild waterfowl. Wildlife Society Bulletin 10:170–172.

World Health Organization. 2005. H5N1 avian influenza: timeline. <www.who.int/csr/disease/avian_ influenza/Timeline_28_10a.pdf> (30 June 2010).

World Health Organization. 2006. Outbreak news: avian influenza, Azerbaijan. Weekly Epidemiology Record 183–188. <www.who.int/wer/en>.

World Health Organization. 2010. Cumulative number of confirmed human cases of avian influenza A/(H5N1) reported to WHO. <www.who .int/csr/disease/avian_influenza/country/cases_ table_2010_05_06/en/index.html> (30 June 2010).

Writing Committee of the World Health Organization Consultation on Human Influenza/A5. 2006. Avian influenza A (H5N1) infection in humans. New England Journal of Medicine 353:1374–1385.

Proper use of DEET and
risks of DEET use

To determine the relative efficacy of DEET versus other insect repellants, Fraidin et al. tested the relative efficacy of seven botanical insect repellents; four products containing N,N-diethyl-m-toluamide, now called N,N-diethyl-3-methylbenzamide (DEET); a repellent containing IR3535 (ethyl butylacetylaminopropionate); three repellent-impregnated wristbands; and a moisturizer that is commonly claimed to have repellent effects. These products were tested in a controlled laboratory environment in which the species of the mosquitoes, their age, their degree of hunger, the humidity, the temperature, and the light–dark cycle were all kept constant.

They found that DEET-based products provided complete protection for the longest duration. Higher concentrations of DEET provided longer-lasting protection. A formulation containing 23.8 percent DEET had a mean complete-protection time of 301.5 minutes. A soybean-oil–based repellent protected against mosquito bites for an average of 94.6 minutes. The IR3535-based repellent protected for an average of 22.9 minutes. All other botanical repellents they tested provided protection for a mean duration of less than 20 minutes. Repellent-impregnated wristbands offered no protection.

They concluded that currently available non-DEET repellents do not provide protection for durations similar to those of DEET-based repellents and cannot be relied on to provide prolonged protection in environments where mosquito-borne diseases are a substantial threat.

Depending on the time in the field, the temperature, exposure to water, perspiration, or concentration of DEET in the product, you may need to re-apply. This study found that a product containing 23.8% DEET provided an average of 5 hours of protection against mosquito bites. A product containing 20% DEET provided almost 4 hours of protection, and a product with 6.65% DEET provided almost 2 hours of protection. DEET may be washed off by perspiration or rain, and its efficacy decreases dramatically with rising outdoor temperatures.

Much has been said about the safety of DEET usage. The Fraidin paper addressed this issue:

Despite the substantial attention paid by the lay press every year to the safety of DEET, this repellent has been subjected to more scientific and toxicologic scrutiny than any other repellent substance. The extensive accumulated toxicologic data on DEET have been reviewed elsewhere. DEET has a remarkable safety profile after 40 years of use and nearly 8 billion human applications. Fewer than 50 cases of serious toxic effects have been documented in the medical literature since 1960, and three quarters of them resolved without sequelae. Many of these cases of toxic effects involved long-term, heavy, frequent, or whole-body application of DEET. No correlation has been found between the concentration of

DEET used and the risk of toxic effects. As part of the Reregistration Eligibility Decision on DEET, released in 1998, the Environmental Protection Agency reviewed the accumulated data on the toxicity of DEET and concluded that "normal use of DEET does not present a health concern to the general U.S. population." When applied with common sense, DEET-based repellents can be expected to provide a safe as well as a long-lasting repellent effect. Until a better repellent becomes available, DEET-based repellents remain the gold standard of protection under circumstances in which it is crucial to be protected against arthropod bites that might transmit disease.

Fradin, M.D., Mark S. and John F. Day, Ph.D. 2002. Comparative Efficacy of Insect Repellents Against Mosquito Bites. New England Journal of Medicine 347: 13–18; available online at <http://content.nejm.org/cgi/content/full/347/1/13>.

INDEX

Kiwi, North Island Brown (*Apteryx australis mantelli*),
malaria prevalence in, 57–60
Kolmogorov-Smirnov test, parasitism in Galápagos
doves, field and lab work, 34–39

laboratory techniques
health precautions, 94–98
parasitism in endemic Galápagos dove, 33–34
trans-Atlantic movement of *Borrelia garinii*, 25–26
Leucocytozoon spp., malaria prevalence in New Zealand
passerines, 56–60
lice, in Galápagos Dove, genetic diversity and immune
response to, 31–39
life history traits, avian hosts and West Nile virus,
seroprevalence patterns and, 13–14
locally weighted nonparametric regression (LOWESS),
parasitism in Galápagos doves, field and lab work,
34–39
logistic regression analysis, avian hosts for West Nile
virus, 7–10
Lyme disease
community complexity and risk of, 5–21
trans-Atlantic movement of *Borrelia garinii* and, 24–28
two-species seroprevalence model for, 15
lymphocyte immunophenotyping, avian disease and, 81–88

Magpie, Black-billed (*Pica pica*), lymphocyte
immunophenotyping of avian hosts and, 85–88
malaria, in endemic New Zealand passerines, 55–60
migratory birds
asymptomatic H5N1 virus in, 70
nondetection of H5N1 avian influenza in, 69–70
pathogenic H5N1 influenza spread by, 68–78
seroprevalence patterns for West Nile virus in, 9–10,
13–14
Mockingbird, Northern (*Mimus polyglottos*), West Nile
virus seroprevalence in, 8, 11
monoclonal antibodies, avian lymphocyte
immunophenotyping, 81–88
morphometric measurements, malaria prevalence in
Bellbirds and, 58–60
mosquito species, avian West Nile virus seroprevalence
and proximity to, 14

nest box monitoring, West Nile virus prevalence and
effects in American kestrel and, 46–51
nesting habits, West Nile virus prevalence and effects in
American kestrel and, 45–52
New Zealand, avian malaria prevalence in, 55–60
North America
forecasting prototype for H5N1 in, 71–78
H5N1 avian influenza in birds in, 71

oceanic islands, trans-Atlantic movement of *Borrelia
garinii* and, 23–28
optical density (OD) values, avian hosts for West Nile
virus, 7
Ornithological Council, zoonotic disease research
by, 91–98
Owl, Great Horned (*Bubo virginianus*), West Nile virus
prevalence in, 51–52

parasitism studies, endemic Galápagos dove, host
genetic diversity and immune response, 31–39
Passeriformes, malaria prevalence in New Zealand
among, 56–60
path analysis, parasitism in endemic Galápagos dove,
35–38
Penguin, King (*Aptenodytes patagonicus*), 25–28
Penguin, Yellow-eyed (*Megadyptes antipodes*), malaria
prevalence in New Zealand in, 57–60
Phthiraptera, parasitism in, 31–39
Physconelloides galapagensis, in Galápagos Dove, genetic
diversity and immune response to, 31–39
Pigeon, Rock (*Columba livia*)
lymphocyte immunophenotyping for WNV
in, 84–88
West Nile virus prevalence in, 48–51
Pipit, New Zealand (*Anthus novaeseelandiae*), malaria
prevalence in, 57–60
Plasmodium relictum capistranoae epizootic, malaria
prevalence in passerines and, 56–60
Plasmodium spp., avian malaria prevalence and, 55–60
predator species, Bellbird extinction and, 57–60
prevalence studies, trans-Atlantic movement of *Borrelia
garinii*, 26–27
principal component analysis, parasitism in Galápagos
doves, field and lab work, 34–39
Puffin, Atlantic (*Fratercula arctica*), *Borrelia garinii*
prevalence in, 23–28
Pyrrhuloxia (*Cardinalis sinuatus*), West Nile virus
seroprevalence in, 8–9

raptors, West Nile virus prevalence and effects
in, 45–52
Raven, Common (*Corvus corax*), 84–88
Razorbills (*Alca torda*), *Borrelia garinii* prevalence in,
23–28
reproductive success
malaria prevalence in Bellbirds and, 60
West Nile virus prevalence and effects in American
kestrel and, 45–52
research methodology
avian hosts for West Nile virus, 5–7
parasitism in endemic Galápagos dove, 33–34
trans-Atlantic movement of *Borrelia garinii*, 25–26
West Nile virus prevalence and effects in American
kestrel, 46–48
zoonotic diseases, safety precautions, 91–98
resident birds, as West Nile virus hosts, 4, 9–10, 14
riparian habitat, avian hosts and West Nile virus
and, 13
Robin, American (*Turdus migratorius*), West Nile virus
seroprevalence in, 8–9, 11, 13–14
Robin, North Island (*Petroica australis*), malaria
prevalence in, 57–60
R-phycoerythrin (R-PE), lymphocyte
immunophenotyping of avian hosts and, 85–88

Saddleback (*Philesturnus carunculatus*), malaria
prevalence in, 57–60
Sage-Grouse, Greater (*Centrocercus urophasianus*), 88
Salmonella spp., banding research and, 92–98

STUDIES IN AVIAN BIOLOGY

Complete Series List

1. Kessel, B., and D. D. Gibson. 1978.
   Status and Distribution of Alaska Birds.

2. Pitelka, F. A., editor. 1979.
   Shorebirds in Marine Environments.

3. Szaro, R. C., and R. P. Balda. 1979.
   Bird Community Dynamics in a Ponderosa Pine Forest.

4. DeSante, D. F., and D. G. Ainley. 1980.
   The Avifauna of the South Farallon Islands, California.

5. Mugaas, J. N., and J. R. King. 1981.
   Annual Variation of Daily Energy Expenditure by the Black-billed Magpie: A Study of Thermal and Behavioral Energetics.

6. Ralph, C. J., and J. M. Scott, editors. 1981.
   Estimating Numbers of Terrestrial Birds.

7. Price, F. E., and C. E. Bock. 1983.
   Population Ecology of the Dipper (Cinclus mexicanus) *in the Front Range of Colorado.*

8. Schreiber, R. W., editor. 1984.
   Tropical Seabird Biology.

9. Scott, J. M., S. Mountainspring, F. L. Ramsey, and C. B. Kepler. 1986.
   Forest Bird Communities of the Hawaiian Islands: Their Dynamics, Ecology, and Conservation.

10. Hand, J. L., W. E. Southern, and K. Vermeer, editors. 1987.
    Ecology and Behavior of Gulls.

11. Briggs, K. T., W. B. Tyler, D. B. Lewis, and D. R. Carlson. 1987.
    Bird Communities at Sea off California: 1975 to 1983.

12. Jehl, J. R., Jr. 1988.
    Biology of the Eared Grebe and Wilson's Phalarope in the Nonbreeding Season: A Study of Adaptations to Saline Lakes.

13. Morrison, M. L., C. J. Ralph, J. Verner, and J. R. Jehl, Jr., editors. 1990.
    Avian Foraging: Theory, Methodology, and Applications.

14. Sealy, S. G., editor. 1990.
    Auks at Sea.

15. Jehl, J. R., Jr., and N. K. Johnson, editors. 1994.
    A Century of Avifaunal Change in Western North America.

16. Block, W. M., M. L. Morrison, and M. H. Reiser, editors. 1994.
    The Northern Goshawk: Ecology and Management.

17. Forsman, E. D., S. DeStefano, M. G. Raphael, and R. J. Gutiérrez, editors. 1996.
    Demography of the Northern Spotted Owl.

18. Morrison, M. L., L. S. Hall, S. K. Robinson, S. I. Rothstein, D. C. Hahn, and T. D. Rich, editors. 1999.
    Research and Management of the Brown-headed Cowbird in Western Landscapes.

19. Vickery, P. D., and J. R. Herkert, editors. 1999.
    Ecology and Conservation of Grassland Birds of the Western Hemisphere.

20. Moore, F. R., editor. 2000.
    Stopover Ecology of Nearctic–Neotropical Landbird Migrants: Habitat Relations and Conservation Implications.

21. Dunning, J. B., Jr., and J. C. Kilgo, editors. 2000.
    Avian Research at the Savannah River Site: A Model for Integrating Basic Research and Long-Term Management.

22. Scott, J. M., S. Conant, and C. van Riper, II, editors. 2001.
    Evolution, Ecology, Conservation, and Management of Hawaiian Birds: A Vanishing Avifauna.

23. Rising, J. D. 2001.
    Geographic Variation in Size and Shape of Savannah Sparrows (Passerculus sandwichensis).

24. Morton, M. L. 2002.
    The Mountain White-crowned Sparrow: Migration and Reproduction at High Altitude.

25. George, T. L., and D. S. Dobkin, editors. 2002.
Effects of Habitat Fragmentation on Birds in Western Landscapes: Contrasts with Paradigms from the Eastern United States.

26. Sogge, M. K., B. E. Kus, S. J. Sferra, and M. J. Whitfield, editors. 2003.
Ecology and Conservation of the Willow Flycatcher.

27. Shuford, W. D., and K. C. Molina, editors. 2004.
Ecology and Conservation of Birds of the Salton Sink: An Endangered Ecosystem.

28. Carmen, W. J. 2004.
Noncooperative Breeding in the California Scrub-Jay.

29. Ralph, C. J., and E. H. Dunn, editors. 2004.
Monitoring Bird Populations Using Mist Nets.

30. Saab, V. A., and H. D. W. Powell, editors. 2005.
Fire and Avian Ecology in North America.

31. Morrison, M. L., editor. 2006.
The Northern Goshawk: A Technical Assessment of Its Status, Ecology, and Management.

32. Greenberg, R., J. E. Maldonado, S. Droege, and M. V. McDonald, editors. 2006.
Terrestrial Vertebrates of Tidal Marshes: Evolution, Ecology, and Conservation.

33. Mason, J. W., G. J. McChesney, W. R. McIver, H. R. Carter, J. Y. Takekawa, R. T. Golightly, J. T. Ackerman, D. L. Orthmeyer, W. M. Perry, J. L. Yee, M. O. Pierson, and M. D. McCrary. 2007.
At-Sea Distribution and Abundance of Seabirds off Southern California: A 20-Year Comparison.

34. Jones, S. L., and G. R. Geupel, editors. 2007.
Beyond Mayfield: Measurements of Nest-Survival Data.

35. Spear, L. B., D. G. Ainley, and W. A. Walker. 2007.
Foraging Dynamics of Seabirds in the Eastern Tropical Pacific Ocean.

36. Niles, L. J., H. P. Sitters, A. D. Dey, P. W. Atkinson, A. J. Baker, K. A. Bennett, R. Carmona, K. E. Clark, N. A. Clark, C. Espoz, P. M. González, B. A. Harrington, D. E. Hernández, K. S. Kalasz, R. G. Lathrop, R. N. Matus, C. D. T. Minton, R. I. G. Morrison, M. K. Peck, W. Pitts, R. A. Robinson, and I. L. Serrano. 2008.
*Status of the Red Knot (*Calidris canutus rufa*) in the Western Hemisphere.*

37. Ruth, J. M., T. Brush, and D. J. Krueper, editors. 2008.
Birds of the US–Mexico Borderland: Distribution, Ecology, and Conservation.

38. Knick, S. T., and J. W. Connelly, editors. 2011.
Greater Sage-Grouse: Ecology and Conservation of a Landscape Species and Its Habitats.

39. Sandercock, B. K., K. Martin, and G. Segelbacher, editors. 2011.
Ecology, Conservation, and Management of Grouse.

40. Forsman, E.D., et al. 2011.
Population Demography of Northern Spotted Owls.

41. Wells, J. V., editor. 2011.
Boreal Birds of North America: A Hemispheric View of Their Conservation Links and Significance.

42. Paul, E., editor. 2012.
Emerging Avian Disease.

Indexer: Diana Witt
Composition: MPS Limited
Text: Scala
Display: Scala Sans
Printer and binder: Thomson-Shore